Aquatic Biodiversity and Water Pollution

Aquatic Biodiversity and Water Pollution

Editor: Ethan Snyder

www.callistoreference.com

Callisto Reference,
118-35 Queens Blvd., Suite 400,
Forest Hills, NY 11375, USA

Visit us on the World Wide Web at:
www.callistoreference.com

ISBN: 978-1-64116-011-7 (Hardback)

Cataloging-in-Publication Data

Aquatic biodiversity and water pollution / edited by Ethan Snyder.
    p. cm.
Includes bibliographical references and index.
ISBN 978-1-64116-011-7
1. Aquatic biodiversity. 2. Water--Pollution. 3. Biodiversity. I. Snyder, Ethan.
QH90.8.B56 A68 2018
577.6--dc23

Table of Contents

Preface

Environment inhabits all the living beings and provides them with means to survive. Today, the biggest threat to the natural environment comes in the form of pollution. Aquatic biodiversity is threatened by water pollution. The ecosystem inside water bodies, consisting of fishes, plants and other water animals gets degraded because of the chemicals, paints, oil spills and waste caused by human beings. This book provides deep insights about aquatic biodiversity and water pollution. For someone with an interest and eye for detail, this book covers the most significant topics in this field.

A detailed account of the significant topics covered in this book is provided below:

Chapter 1- An aquatic ecosystem refers to the symbiotic relationship of organisms with non-living elements in a water body. The major types of ecosystems are marine ecosystem, freshwater ecosystem and lake ecosystem. They are important for the environment as they purify water, replenish ground water, provide shelter to many flora and fauna, etc. This chapter will provide an integrated understanding of aquatic ecosystem.

Chapter 2- Aquatic biodiversity is seen as the amount of living to nonliving matter in an aquatic region. 'Hotspots' are regions that possess a large amount of fish species. Aquatic biodiversity can be found in both freshwater and seawater environment. This section is an overview of the subject matter incorporating all the major aspects of aquaculture.

Chapter 3- Population explosion, accompanied with urbanization and industrialization, has led to a hazardous impact on aquatic life. With the human populace indulging in overfishing, the food chain can become disturbed, and cause extinction or overpopulation of fish species. This section has been carefully written to provide an easy understanding of the varied facets of human impact on aquatic biodiversity.

Chapter 4- Important areas of marine conservation include habitat alteration and loss, species introduction of aquatic organisms, organisms facing extinction, etc. Examples of marine ecosystem management in different regions have been discussed. The aspects elucidated in this chapter are of vital importance, and provide a better understanding of marine conservation and management.

It gives me an immense pleasure to thank our entire team for their efforts. Finally in the end, I would like to thank my family and colleagues who have been a great source of inspiration and support.

Editor

Understanding Aquatic Ecosystem

An aquatic ecosystem refers to the symbiotic relationship of organisms with non-living elements in a water body. The major types of ecosystems are marine ecosystem, freshwater ecosystem and lake ecosystem. They are important for the environment as they purify water, replenish ground water, provide shelter to many flora and fauna, etc. This chapter will provide an integrated understanding of aquatic ecosystem.

Aquatic Ecosystem

An aquatic ecosystem is an ecosystem in a body of water. Communities of organisms that are dependent on each other and on their environment live in aquatic ecosystems. The two main types of aquatic ecosystems are marine ecosystems and freshwater ecosystems.

An estuary mouth and coastal waters, part of an aquatic ecosystem

Types

Marine

Marine ecosystems cover approximately 71% of the Earth's surface and contain approximately 97% of the planet's water. They generate 32% of the world's net primary production. They are distinguished from freshwater ecosystems by the presence of dissolved compounds, especially salts, in the water. Approximately 85% of the dissolved materials in seawater are sodium and chlorine. Seawater has an average salinity of 35 parts per thousand (ppt) of water. Actual salinity varies among different marine ecosystems.

Marine ecosystems can be divided into many zones depending upon water depth and shoreline features. The oceanic zone is the vast open part of the ocean where animals such as whales, sharks, and tuna live. The benthic zone consists of substrates below water where many invertebrates live. The intertidal zone is the area between high and low tides; in this figure it is termed the littoral zone. Other near-shore (neritic) zones can include estuaries, salt marshes, coral reefs, lagoons and mangrove swamps. In the deep water, hydrothermal vents may occur where chemosynthetic sulfur bacteria form the base of the food web.

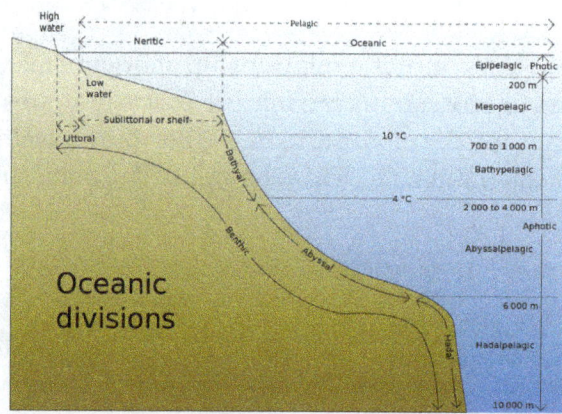

A classification of marine habitats.

Classes of organisms found in marine ecosystems include brown algae, dinoflagellates, corals, cephalopods, echinoderms, and sharks. Fishes caught in marine ecosystems are the biggest source of commercial foods obtained from wild populations.

Environmental problems concerning marine ecosystems include unsustainable exploitation of marine resources (for example overfishing of certain species), marine pollution, climate change, and building on coastal areas.

Freshwater

Freshwater ecosystem.

Freshwater ecosystems cover 0.78% of the Earth's surface and inhabit 0.009% of its total water. They generate nearly 3% of its net primary production. Freshwater ecosystems contain 41% of the world's known fish species.

There are three basic types of freshwater ecosystems:

- Lentic: slow moving water, including pools, ponds, and lakes.

- Lotic: faster moving water, for example streams and rivers.

- Wetlands: areas where the soil is saturated or inundated for at least part of the time.

Lentic

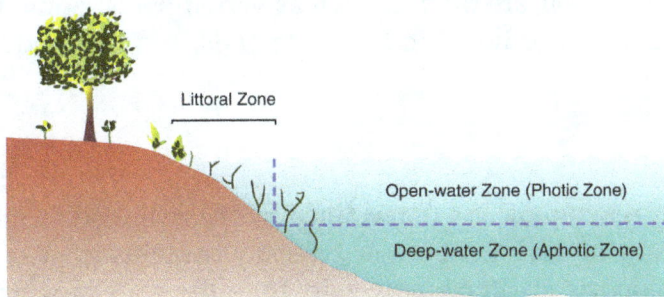

The three primary zones of a lake.

Lake ecosystems can be divided into zones. One common system divides lakes into three zones. The first, the littoral zone, is the shallow zone near the shore. This is where rooted wetland plants occur. The offshore is divided into two further zones, an open water zone and a deep water zone. In the open water zone (or photic zone) sunlight supports photosynthetic algae, and the species that feed upon them. In the deep water zone, sunlight is not available and the food web is based on detritus entering from the littoral and photic zones. Some systems use other names. The off shore areas may be called the pelagic zone, the photic zone may be called the limnetic zone and the aphotic zone may be called the profundal zone. Inland from the littoral zone one can also frequently identify a riparian zone which has plants still affected by the presence of the lake—this can include effects from windfalls, spring flooding, and winter ice damage. The production of the lake as a whole is the result of production from plants growing in the littoral zone, combined with production from plankton growing in the open water.

Wetlands can be part of the lentic system, as they form naturally along most lake shores, the width of the wetland and littoral zone being dependent upon the slope of the shoreline and the amount of natural change in water levels, within and among years. Often dead trees accumulate in this zone, either from windfalls on the shore or logs transported to the site during floods. This woody debris provides important habitat for fish and nesting birds, as well as protecting shorelines from erosion.

Two important subclasses of lakes are ponds, which typically are small lakes that intergrade with wetlands, and water reservoirs. Over long periods of time, lakes, or bays within them, may gradually become enriched by nutrients and slowly fill in with organic sediments, a process called succession. When humans use the watershed, the volumes of sediment entering the lake can accelerate this process. The addition of sediments and nutrients to a lake is known as eutrophication.

Ponds

Ponds are small bodies of freshwater with shallow and still water, marsh, and aquatic plants. They can be further divided into four zones: vegetation zone, open water, bottom mud and surface film. The size and depth of ponds often varies greatly with the time of year; many ponds are produced by spring flooding from rivers. Food webs are based both on free-floating algae and upon aquatic plants. There is usually a diverse array of aquatic life, with a few examples including algae, snails, fish, beetles, water bugs, frogs, turtles, otters and muskrats. Top predators may include large fish, herons, or alligators. Since fish are a major predator upon amphibian larvae, ponds that dry up each year, thereby killing resident fish, provide important refugia for amphibian breeding. Ponds that dry up completely each year are often known as vernal pools. Some ponds are produced by animal activity, including alligator holes and beaver ponds, and these add important diversity to landscapes.

Lotic

The major zones in river ecosystems are determined by the river bed's gradient or by the velocity of the current. Faster moving turbulent water typically contains greater concentrations of dissolved oxygen, which supports greater biodiversity than the slow moving water of pools. These distinctions form the basis for the division of rivers into upland and lowland rivers. The food base of streams within riparian forests is mostly derived from the trees, but wider streams and those that lack a canopy derive the majority of their food base from algae. Anadromous fish are also an important source of nutrients. Environmental threats to rivers include loss of water, dams, chemical pollution and introduced species. A dam produces negative effects that continue down the watershed. The most important negative effects are the reduction of spring flooding, which damages wetlands, and the retention of sediment, which leads to loss of deltaic wetlands.

Wetlands

Wetlands are dominated by vascular plants that have adapted to saturated soil. There are four main types of wetlands: swamp, marsh, fen and bog (both fens and bogs are types of mire). Wetlands are the most productive natural ecosystems in the world because of the proximity of water and soil. Hence they support large numbers of plant and animal species. Due to their productivity, wetlands are often converted into dry land with dykes and drains and used for agricultural purposes. The construction of dykes, and dams, has negative consequences for individual wetlands and entire watersheds. Their closeness to lakes and rivers means that they are often developed for human settlement. Once settlements are constructed and protected by dykes, the settlements then become vulnerable to land subsidence and ever increasing risk of flooding. The Louisiana coast around New Orleans is a well-known example; the Danube Delta in Europe is another.

Functions

Aquatic ecosystems perform many important environmental functions. For example, they recycle nutrients, purify water, attenuate floods, recharge ground water and provide habitats for wildlife. Aquatic ecosystems are also used for human recreation, and are very important to the tourism industry, especially in coastal regions.

The health of an aquatic ecosystem is degraded when the ecosystem's ability to absorb a stress has been exceeded. A stress on an aquatic ecosystem can be a result of physical, chemical or biological alterations of the environment. Physical alterations include changes in water temperature, water flow and light availability. Chemical alterations include changes in the loading rates of biostimulatory nutrients, oxygen consuming materials, and toxins. Biological alterations include over-harvesting of commercial species and the introduction of exotic species. Human populations can impose excessive stresses on aquatic ecosystems. There are many examples of excessive stresses with negative consequences. Consider three. The environmental history of the Great Lakes of North America illustrates this problem, particularly how multiple stresses, such as water pollution, over-harvesting and invasive species can combine. The Norfolk Broadlands in England illustrate similar decline with pollution and invasive species. Lake Pontchartrain along the Gulf of Mexico illustrates the negative effects of different stresses including levee construction, logging of swamps, invasive species and salt water intrusion.

Abiotic Characteristics

An ecosystem is composed of biotic communities that are structured by biological interactions and abiotic environmental factors. Some of the important abiotic environmental factors of aquatic ecosystems include substrate type, water depth, nutrient levels, temperature, salinity, and flow. It is often difficult to determine the relative importance of these factors without rather large experiments. There may be complicated feedback loops. For example, sediment may determine the presence of aquatic plants, but aquatic plants may also trap sediment, and add to the sediment through peat.

The amount of dissolved oxygen in a water body is frequently the key substance in determining the extent and kinds of organic life in the water body. Fish need dissolved oxygen to survive, although their tolerance to low oxygen varies among species; in extreme cases of low oxygen some fish even resort to air gulping. Plants often have to produce aerenchyma, while the shape and size of leaves may also be altered. Conversely, oxygen is fatal to many kinds of anaerobic bacteria.

Nutrient levels are important in controlling the abundance of many species of algae. The relative abundance of nitrogen and phosphorus can in effect determine which species of algae come to dominate. Algae are a very important source of food for aquatic life, but at the same time, if they become over-abundant, they can cause declines in fish when they decay. Similar over-abundance of algae in coastal environments such as the Gulf of Mexico produces, upon decay, a hypoxic region of water known as a dead zone.

The salinity of the water body is also a determining factor in the kinds of species found in the water body. Organisms in marine ecosystems tolerate salinity, while many freshwater organisms are intolerant of salt. The degree of salinity in an estuary or delta is an important control upon the type of wetland (fresh, intermediate, or brackish), and the associated animal species. Dams built upstream may reduce spring flooding, and reduce sediment accretion, and may therefore lead to saltwater intrusion in coastal wetlands.

Freshwater used for irrigation purposes often absorbs levels of salt that are harmful to freshwater organisms.

Biotic Characteristics

The biotic characteristics are mainly determined by the organisms that occur. For example, wetland plants may produce dense canopies that cover large areas of sediment—or snails or geese may graze the vegetation leaving large mud flats. Aquatic environments have relatively low oxygen levels, forcing adaptation by the organisms found there. For example, many wetland plants must produce aerenchyma to carry oxygen to roots. Other biotic characteristics are more subtle and difficult to measure, such as the relative importance of competition, mutualism or predation. There are a growing number of cases where predation by coastal herbivores including snails, geese and mammals appears to be a dominant biotic factor.

Autotrophic Organisms

Autotrophic organisms are producers that generate organic compounds from inorganic material. Algae use solar energy to generate biomass from carbon dioxide and are possibly the most important autotrophic organisms in aquatic environments. Of course, the more shallow the water, the greater the biomass contribution from rooted and floating vascular plants. These two sources combine to produce the extraordinary production of estuaries and wetlands, as this autotrophic biomass is converted into fish, birds, amphibians and other aquatic species.

Chemosynthetic bacteria are found in benthic marine ecosystems. These organisms are able to feed on hydrogen sulfide in water that comes from volcanic vents. Great concentrations of animals that feed on these bacteria are found around volcanic vents. For example, there are giant tube worms (*Riftia pachyptila*) 1.5 m in length and clams (*Calyptogena magnifica*) 30 cm long.

Heterotrophic Organisms

Heterotrophic organisms consume autotrophic organisms and use the organic compounds in their bodies as energy sources and as raw materials to create their own biomass. Euryhaline organisms are salt tolerant and can survive in marine ecosystems, while stenohaline or salt intolerant species can only live in freshwater environments.

Aquatic ecosystem is the most diverse ecosystem in the world. The first life originated in the water and first organisms were also aquatic where water was the principal external as well as internal medium for organisms. Thus water is the most vital factor for the existence of all living organisms. Water covers about 71% of the earth of which more than 95% exists in gigantic oceans. A very less amount of water is contained in the rivers (0.00015%) and lakes (0.01%), which comprise the most valuable fresh water resources. Global aquatic ecosystems fall under two broad classes defined by salinity – freshwater ecosystem and the saltwater ecosystem. Freshwater ecosystems are inland waters that have low concentrations of salts (< 500 mg/L). The salt-water ecosystem has high concentration of salt content (averaging about 3.5%).

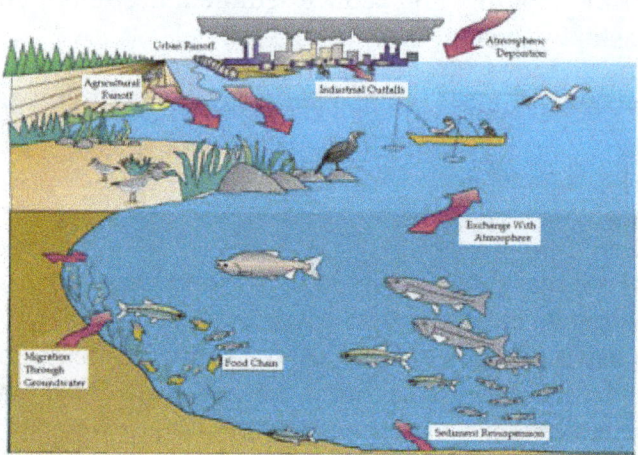

An aquatic ecosystem (habitats and organisms) includes rivers and streams, ponds and lakes, oceans and bays, and swamps and marshes, and their associated animals. These species have evolved and adapted to watery habitats over millions of years. Aquatic habitats provide the food, water, shelter, and space essential for the survival of aquatic animals and plants. Aquatic biodiversity is the rich and harbors variety of plants and animals-from primary producers algae to tertiary consumers large fishes, intermittently occupied by zooplankton, small fishes, aquatic insects and amphibians. Many of these animals and plants species live in water; some like fish spend all their lives underwater, whereas others, like toads and frogs, may use surface waters only during the breeding season or as juveniles.

Freshwater Ecosystem

Freshwater ecosystems cover 0.8% of the Earth's surface and contain 0.009% of its total water. They generate nearly 3% of its net primary production. Freshwater ecosystes contain 41% of the world's known fish species.

The study of freshwater habitats is known as limnology. Freshwater habitats can be further divided into two groups as lentic and lotic ecosystems based on the difference in the water residence time and the flow velocity. The water residence time in a lentic ecosystem on an average is 10 years and that of lotic ecosystem is 2 weeks. In lotic ecosystem, the average flow velocity ranges from 0.1 to 1 m/s whereas lentic ecosystems are characterized by an average flow velocity of 0.001 to 0.01 m/s (Wetzel, 2001; UNEP, 1996). The lentic habitats further differentiate from lotic habitats by having a thermal stratification with is created in a lake due to differences in densities. Water reaches a maximum density at 40C, a warm, lighter water floats on top of the heavier cooler water thus creating thermally stratified zones which corresponds to epilimnion, the warm layer, the hypolimnion, the colder layer separated by a barrier called thermocline. The lotic ecosystem is characterized by stream orders depending on the origin and flow and various types of stream pattern namely Dendritic, Radial, Rectangular, Centripetal, Pinnate, Trellis, Parallel, Distributory and Annular, which determines the flooding and soil erosion hazards of the region. However, the basic unity among these ecosystems is that any alteration in the catchment area of these ecosystems will affect the water quality of both lotic and lentic ecosystem. The catchment area is all land and water area, which contributes runoff to a common point, which may be a lake or a stream. The term catchment is equivalent to drainage basin and watershed (Davie, 2002; Tideman, 2000).

The term lotic (from lavo, meaning 'to wash') represents running water, where the entire body of water moves in a definite direction. It includes spring, stream, or river viewed as an ecological unit of the biotic community and the physiochemical environment. Lotic ecosystems are characterized by the interaction between flowing water with a longitudinal gradation in temperature, organic and inorganic materials, energy, and the organisms within a stream corridor. These interactions occur over space and time.

Biological Characteristics:

Lentic Ecosystem

The biological characteristics of still water bodies may be broadly classified into – pelagic and benthic systems. Benthic system is subdivided into littoral and profundal types. The species composition of communities of all those types is greatly influenced by the nutrient status of the water concerned. The pelagic habitat is that of the open water away from the influence of shore or bottom substrate, while benthic habitat is associated with the substrate of the lake. The littoral habitat is extending from the shoreline out to the deeper water. The plankton community, phytoplankton and zooplankton, occupy the regions of high light intensities namely on the surface layer of pelagic zone and the littoral zone. Some of the zooplankton members also inhabit the benthic zone feeding on detritus and sinking phytoplankton. Fishes occupy the littoral, pelagic and occasionally profundal zones, when the dissolved oxygen content in the lake is high. Macroinvertebrates are confined to the benthic zone.

Marine Ecosystem

Marine ecosystems are among the largest of Earth's aquatic ecosystems. Examples include salt marshes, intertidal zones, estuaries, lagoons, mangroves, coral reefs, the deep sea, and the sea floor. They can be contrasted with freshwater ecosystems, which have a lower salt content. Marine waters cover two-thirds of the surface of the Earth. Such places are considered ecosystems because the plant life supports the animal life and vice versa.

Marine ecosystems are essential for the overall health of both marine and terrestrial environments. According to the World Resource Center, coastal habitats account for about one-third of marine biological productivity. Estuarine ecosystems, such as salt marshes, seagrass meadows and mangrove forests, are among the most productive ecosystems on the planet. Coral reefs provide food and shelter to the highest levels of marine diversity in the world.

Coral reefs form complex marine ecosystems with tremendous biodiversity.

Here, we can see different types of starfish, coral reefs and fishes in the Great Barrier Reef.

Marine ecosystems usually have a large biodiversity and are therefore thought to have a good resistance against invasive species. However, exceptions have been observed, and the mechanisms responsible in determining the success of an invasion are not yet clear.

Freshwater Ecosystem

Freshwater ecosystems are a subset of Earth's aquatic ecosystems. They include lakes and ponds, rivers, streams, springs, and wetlands. They can be contrasted with marine ecosystems, which have a larger salt content. Freshwater habitats can be classified by different factors, including temperature, light penetration, and vegetation.

Freshwater ecosystems can be divided into lentic ecosystems (still water) and lotic ecosystems (flowing water).

Limnology (and its branch freshwater biology) is a study about freshwater ecosystems. It is a part of hydrobiology.

Original attempts to understand and monitor freshwater ecosystems were spurred on by threats to human health (ex. Cholera outbreaks due to sewage contamination). Early monitoring focussed on chemical indicators, then bacteria, and finally algae, fungi and protozoa. A new type of monitoring involves differing groups of organisms (macroinvertebrates, macrophytes and fish) and the stream conditions associated with them.

Current biomonitoring techniques focus mainly on community structure or biochemical oxygen demand. Responses are measured by behavioural changes, altered rates of growth, reproduction or mortality. Macroinvertebrates are most often used in these models because of well known taxonomy, ease of collection, sensitivity to a range of stressors, and their overall value to the ecosystem. Most of these measurements are difficult to extrapolate on a large scale, however.

The use of reference sites is common when assessing what a healthy freshwater ecosystem should "look like". Reference sites are easier to reconstruct in standing water than moving water. Preserved indicators such as diatom valves, macrophyte pollen, insect chitin and fish scales can be used to establish a reference ecosystem representative of a time before large scale human disturbance.

Common chemical stresses on freshwater ecosystem health include acidification, eutrophication and copper and pesticide contamination.

Extinction of Freshwater Fauna

Over 123 freshwater fauna species have gone extinct in North America since 1900. Of North American freshwater species, an estimated 48.5% of mussels, 22.8% of gastropods, 32.7% of crayfishes, 25.9% of amphibians, and 21.3% of fish are either endangered or threatened. Extinction rates of many species may increase severely into the next century because of invasive species, loss of keystone species, and species which are already functionally extinct. Projected extinction rates for freshwater animals are around five times greater than for land animals, and are comparable to the rates for rainforest communities. Recent extinction trends can be attributed largely to sedimentation, stream fragmentation, chemical and organic pollutants, dams, and invasive species.

Lake Ecosystem

A lake ecosystem includes biotic (living) plants, animals and micro-organisms, as well as abiotic (nonliving) physical and chemical interactions.

Lake ecosystems are a prime example of lentic ecosystems. *Lentic* refers to stationary or relatively still water, from the Latin *lentus*, which means sluggish. Lentic ecosystems can be compared with lotic ecosystems, which involve flowing terrestrial waters such as rivers and streams. Together, these two fields form the more general study area of freshwater or aquatic ecology.

Lentic systems are diverse, ranging from a small, temporary rainwater pool a few inches deep to Lake Baikal, which has a maximum depth of 1740 m. The general distinction between pools/ponds and lakes is vague, but Brown states that ponds and pools have their entire bottom surfaces exposed to light, while lakes do not. In addition, some lakes become seasonally stratified. Ponds and pools have two regions: the pelagic open water zone, and the benthic zone, which comprises the bottom and shore regions. Since lakes have deep bottom regions not exposed to light, these systems have an additional zone, the profundal. These three areas can have very different abiotic conditions and, hence, host species that are specifically adapted to live there.

Important Abiotic Factors

Light

Light provides the solar energy required to drive the process of photosynthesis, the major energy source of lentic systems. The amount of light received depends upon a combination of several factors. Small ponds may experience shading by surrounding trees, while cloud cover may affect light availability in all systems, regardless of size. Seasonal and diurnal considerations also play a role in light availability because the shallower the angle at which light strikes water, the more light is lost by reflection. This is known as Beer's law. Once light has penetrated the surface, it may also be scattered by particles suspended in the water column. This scattering decreases the total amount of light as depth increases. Lakes are divided into photic and aphotic regions, the prior receiving sunlight and latter being below the depths of light penetration, making it void of photosynthetic capacity. In relation to lake zonation, the pelagic and benthic zones are considered to lie within the photic region, while the profundal zone is in the aphotic region.

Temperature

Temperature is an important abiotic factor in lentic ecosystems because most of the biota are poikilothermic, where internal body temperatures are defined by the surrounding system. Water can be heated or cooled through radiation at the surface and conduction to or from the air and surrounding substrate. Shallow ponds often have a continuous temperature gradient from warmer waters at the surface to cooler waters at the bottom. In addition, temperature fluctuations can be very great in these systems, both diurnally and seasonally.

Temperature regimes are very different in large lakes. In temperate regions, for example, as air temperatures increase, the icy layer formed on the surface of the lake breaks up, leaving the water at approximately 4 °C. This is the temperature at which water has the highest density. As the season progresses, the warmer air temperatures heat the surface waters, making them less dense. The deeper waters remain cool and dense due to reduced light penetration. As the summer begins, two distinct layers become established, with such a large temperature difference between them that they remain stratified. The lowest zone in the lake is the coldest and is called the hypolimnion. The upper warm zone is called the epilimnion. Between these zones is a band of rapid temperature change called the thermocline. During the colder fall season, heat is lost at the surface and the epilimnion cools. When the temperatures of the two zones are close enough, the waters begin to mix again to create a uniform temperature, an event termed lake turnover. In the winter, inverse stratification occurs as water near the surface cools freezes, while warmer, but denser water remains near the bottom. A thermocline is established, and the cycle repeats.

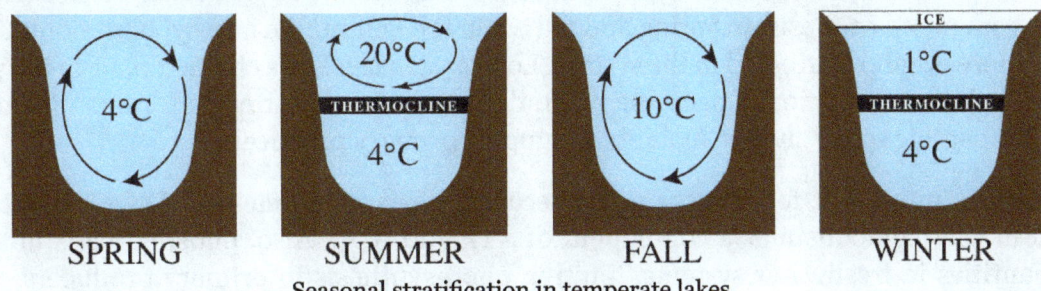

Seasonal stratification in temperate lakes

Wind

In exposed systems, wind can create turbulent, spiral-formed surface currents called Langmuir circulations. Exactly how these currents become established is still not well understood, but it is evident that it involves some interaction between horizontal surface currents and surface gravity waves. The visible result of these rotations, which can be seen in any lake, are the surface foamlines that run parallel to the wind direction. Positively buoyant particles and small organisms concentrate in the foamline at the surface and negatively buoyant objects are found in the upwelling current between the two rotations. Objects with neutral buoyancy tend to be evenly distributed in the water column. This turbulence circulates nutrients in the water column, making it crucial for many pelagic species, however its effect on benthic and profundal organisms is minimal to non-existent, respectively. The degree of nutrient circulation is system specific, as it depends upon such factors as wind strength and duration, as well as lake or pool depth and productivity.

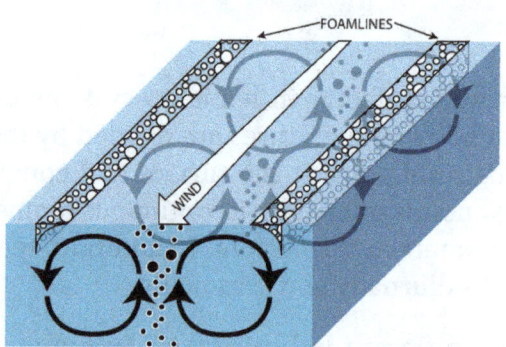

Illustration of Langmuir rotations; open circles=positively buoyant particles, closed circles=negatively buoyant particles

Chemistry

Oxygen is essential for organismal respiration. The amount of oxygen present in standing waters depends upon: 1) the area of transparent water exposed to the air, 2) the circulation of water within the system and 3) the amount of oxygen generated and used by organisms present. In shallow, plant-rich pools there may be great fluctuations of oxygen, with extremely high concentrations occurring during the day due to photosynthesis and very low values at night when respiration is the dominant process of primary producers. Thermal stratification in larger systems can also affect the amount of oxygen present in different zones. The epilimnion is oxygen rich because it circulates quickly, gaining oxygen via contact with the air. The hypolimnion, however, circulates very slowly and has no atmospheric contact. Additionally, fewer green plants exist in the hypolimnion, so there is less oxygen released from photosynthesis. In spring and fall when the epilimnion and hypolimnion mix, oxygen becomes more evenly distributed in the system. Low oxygen levels are characteristic of the profundal zone due to the accumulation of decaying vegetation and animal matter that "rains" down from the pelagic and benthic zones and the inability to support primary producers.

Phosphorus is important for all organisms because it is a component of DNA and RNA and is involved in cell metabolism as a component of ATP and ADP. Also, phosphorus is not found in large quantities in freshwater systems, limiting photosynthesis in primary producers, making it the main determinant of lentic system production. The phosphorus cycle is complex, but the model outlined below describes the basic pathways. Phosphorus mainly enters a pond or lake through runoff from the watershed or by atmospheric deposition. Upon entering the system, a reactive form of phosphorus is usually taken up by algae and macrophytes, which release a non-reactive phosphorus compound as a byproduct of photosynthesis. This phosphorus can drift downwards and become part of the benthic or profundal sediment, or it can be remineralized to the reactive form by microbes in the water column. Similarly, non-reactive phosphorus in the sediment can be remineralized into the reactive form. Sediments are generally richer in phosphorus than lake water, however, indicating that this nutrient may have a long residency time there before it is remineralized and re-introduced to the system.

Lentic System Biota

Bacteria

Bacteria are present in all regions of lentic waters. Free-living forms are associated with decomposing organic material, biofilm on the surfaces of rocks and plants, suspended in the water column, and in the sediments of the benthic and profundal zones. Other forms are also associated with the guts of lentic animals as parasites or in commensal relationships. Bacteria play an important role in system metabolism through nutrient recycling, which is discussed in the Trophic Relationships section.

Primary Producers

Nelumbo nucifera, an aquatic plant.

Algae, including both phytoplankton and periphyton are the principle photosynthesizers in ponds and lakes. Phytoplankton are found drifting in the water column of the pelagic zone. Many species have a higher density than water which should make them sink and end up in the benthos. To combat this, phytoplankton have developed density changing mechanisms, by forming vacuoles and gas vesicles or by changing their shapes to induce drag, slowing their descent. A very sophisticated adaptation utilized by a small number of species is a tail-like flagellum that can adjust vertical position and allow movement in any direction. Phytoplankton can also maintain their presence in the water column by being circulated in Langmuir rotations. Periphytic algae, on the other hand, are attached to a substrate. In lakes and ponds, they can cover all benthic surfaces. Both types of plankton are important as food sources and as oxygen providers.

Aquatic plants live in both the benthic and pelagic zones and can be grouped according to their manner of growth: 1) emergent = rooted in the substrate but with leaves and flowers extending into the air, 2) floating-leaved = rooted in the substrate but with floating leaves, 3) submersed = growing beneath the surface and 4) free-floating macrophytes = not rooted in the substrate and floating on the surface. These various forms of macrophytes generally occur in different areas of the benthic zone, with emergent vegetation nearest the shoreline, then floating-leaved macrophytes, followed by submersed vegetation. Free-floating macrophytes can occur anywhere on the system's surface.

Aquatic plants are more buoyant than their terrestrial counterparts because freshwater has a higher density than air. This makes structural rigidity unimportant in lakes and ponds (except in the aerial stems and leaves). Thus, the leaves and stems of most aquatic plants use less energy to construct and maintain woody tissue, investing that energy into fast growth instead. In order to contend with stresses induced by wind and waves, plants must be both flexible and tough. Light, water depth and substrate types are the most important factors controlling the distribution of submerged aquatic plants. Macrophytes are sources of food, oxygen, and habitat structure in the benthic zone, but cannot penetrate the depths of the euphotic zone and hence are not found there.

Invertebrates

Water striders are predatory insects which rely on surface tension to walk on top of water. They live on the surface of ponds, marshes, and other quiet waters. They can move very quickly, up to 1.5 m/s.

Zooplankton are tiny animals suspended in the water column. Like phytoplankton, these species have developed mechanisms that keep them from sinking to deeper waters, including drag-inducing body forms and the active flicking of appendages such as antennae or spines. Remaining in the water column may have its advantages in terms of feeding, but this zone's lack of refugia leaves zooplankton vulnerable to predation. In response, some species, especially Daphnia sp., make daily vertical migrations in the water column by passively sinking to the darker lower depths during the day and actively moving towards the surface during the night. Also, because conditions in a lentic system can be quite variable across seasons, zooplankton have the ability to switch from laying regular eggs to resting eggs when there is a lack of food, temperatures fall below 2 °C, or if predator abundance is high. These resting eggs have a diapause, or dormancy period that should allow the zooplankton to encounter conditions that are more favorable to survival when they finally hatch. The invertebrates that inhabit the benthic zone are numerically dominated by small species and are species rich compared to the zooplankton of the open water. They include Crustaceans (e.g. crabs, crayfish, and shrimp), molluscs (e.g. clams and snails), and numerous types of insects. These organisms are mostly found in the areas of macrophyte growth, where the richest resources, highly oxygenated water, and warmest portion of the ecosystem are found. The structurally diverse macrophyte beds are important sites for the accumulation of organic matter, and provide an ideal area for colonization. The sediments and plants also offer a great deal of protection from predatory fishes.

Very few invertebrates are able to inhabit the cold, dark, and oxygen poor profundal zone. Those that can are often red in color due to the presence of large amounts of hemoglobin, which greatly increases the amount of oxygen carried to cells. Because the concentration of oxygen within this zone is low, most species construct tunnels or borrows in which they can hide and make the minimum movements necessary to circulate water through, drawing oxygen to them without expending much energy.

Fish and other Vertebrates

Fish have a range of physiological tolerances that are dependent upon which species they belong to. They have different lethal temperatures, dissolved oxygen requirements, and spawning needs that are based on their activity levels and behaviors. Because fish are highly mobile, they are able to deal with unsuitable abiotic factors in one zone by simply moving to another. A detrital feeder in the profundal zone, for example, that finds the oxygen concentration has dropped too low may feed closer to the benthic zone. A fish might also alter its residence during different parts of its life history: hatching in a sediment nest, then moving to the weedy benthic zone to develop in a protected environment with food resources, and finally into the pelagic zone as an adult.

Other vertebrate taxa inhabit lentic systems as well. These include amphibians (e.g. salamanders and frogs), reptiles (e.g. snakes, turtles, and alligators), and a large number of waterfowl species. Most of these vertebrates spend part of their time in terrestrial habitats and thus are not directly affected by abiotic factors in the lake or pond. Many fish species are important as consumers and as prey species to the larger vertebrates.

Trophic Relationships

Primary Producers

Lentic systems gain most of their energy from photosynthesis performed by aquatic plants and algae. This autochthonous process involves the combination of carbon dioxide, water, and solar energy to produce carbohydrates and dissolved oxygen. Within a lake or pond, the potential rate of photosynthesis generally decreases with depth due to light attenuation. Photosynthesis, however, is often low at the top few millimeters of the surface, likely due to inhibition by ultraviolet light. The exact depth and photosynthetic rate measurements of this curve are system specific and depend upon: 1) the total biomass of photosynthesizing cells, 2) the amount of light attenuating materials and 3) the abundance and frequency range of light absorbing pigments (i.e. chlorophylls) inside of photosynthesizing cells. The energy created by these primary producers is important for the community because it is transferred to higher trophic levels via consumption.

Bacteria

The vast majority of bacteria in lakes and ponds obtain their energy by decomposing vegetation and animal matter. In the pelagic zone, dead fish and the occasional allochthonous input of litterfall are examples of coarse particulate organic matter (CPOM>1 mm). Bacteria degrade these into fine particulate organic matter (FPOM<1 mm) and then further into usable nutrients. Small organisms such as plankton are also characterized as FPOM. Very low concentrations of nutrients are released during decomposition because the bacteria are utilizing them to build their own biomass. Bacteria, however, are consumed by protozoa, which are in turn consumed by zooplankton, and then further up the trophic levels. Nutrients, including those that contain carbon and phosphorus, are reintroduced into the water column at any number of points along this food chain via excretion or organism death, making them available again for bacteria. This regeneration cycle is known as the microbial loop and is a key component of lentic food webs.

The decomposition of organic materials can continue in the benthic and profundal zones if the matter falls through the water column before being completely digested by the pelagic bacteria. Bacteria are found in the greatest abundance here in sediments, where they are typically 2-1000 times more prevalent than in the water column.

Benthic Invertebrates

Benthic invertebrates, due to their high level of species richness, have many methods of prey capture. Filter feeders create currents via siphons or beating cilia, to pull water and its nutritional contents, towards themselves for straining. Grazers use scraping, rasping, and shredding adaptations to feed on periphytic algae and macrophytes. Members of the collector guild browse the sediments, picking out specific particles with raptorial appendages. Deposit feeding invertebrates indiscriminately consume sediment, digesting any organic material it contains. Finally, some invertebrates belong to the predator guild, capturing and consuming living animals. The profundal zone is home to a unique group of filter feeders that use small body movements to draw a current through burrows that they have created in the sediment. This mode of feeding requires the least amount of motion, allowing these species to conserve energy. A small number of invertebrate taxa are predators in the profundal zone. These species are likely from other regions and only come to these depths to feed. The vast majority of invertebrates in this zone are deposit feeders, getting their energy from the surrounding sediments.

Fish

Fish size, mobility, and sensory capabilities allow them to exploit a broad prey base, covering multiple zonation regions. Like invertebrates, fish feeding habits can be categorized into guilds. In the pelagic zone, herbivores graze on periphyton and macrophytes or pick phytoplankton out of the water column. Carnivores include fishes that feed on zooplankton in the water column (zooplanktivores), insects at the water's surface, on benthic structures, or in the sediment (insectivores), and those that feed on other fish (piscivores). Fish that consume detritus and gain energy by processing its organic material are called detritivores. Omnivores ingest a wide variety of prey, encompassing floral, faunal, and detrital material. Finally, members of the parasitic guild acquire nutrition from a host species, usually another fish or large vertebrate. Fish taxa are flexible in their feeding roles, varying their diets with environmental conditions and prey availability. Many species also undergo a diet shift as they develop. Therefore, it is likely that any single fish occupies multiple feeding guilds within its lifetime.

Lentic Food Webs

The lentic biota are linked in complex web of trophic relationships. These organisms can be considered to loosely be associated with specific trophic groups (e.g. primary producers, herbivores, primary carnivores, secondary carnivores, etc.). Scientists have developed several theories in order to understand the mechanisms that control the abundance and diversity within these groups. Very generally, top-down processes dictate that the abundance of prey taxa is dependent upon the actions of consumers from higher trophic levels. Typically, these processes operate only between two trophic levels, with no effect on the others. In some cases, however, aquatic systems experience a trophic cascade; for example, this might occur if primary producers experience less grazing by herbivores because these herbivores are suppressed by carnivores. Bottom-up processes are functioning when the abundance or diversity of members of higher trophic levels is dependent upon the availability or quality of resources from lower levels. Finally, a combined regulating theory, bottom-up:top-down, combines the predicted influences of consumers and resource availability. It predicts that trophic levels close to the lowest trophic levels will be most influenced by bottom-up forces, while top-down effects should be strongest at top levels.

Community Patterns and Diversity

Local Species Richness

The biodiversity of a lentic system increases with the surface area of the lake or pond. This is attributable to the higher likelihood of partly terrestrial species of finding a larger system. Also, because larger systems typically have larger populations, the chance of extinction is decreased. Additional factors, including temperature regime, pH, nutrient availability, habitat complexity, speciation rates, competition, and predation, have been linked to the number of species present within systems.

Succession Patterns in Plankton Communities – the PEG Model

Phytoplankton and zooplankton communities in lake systems undergo seasonal succession in relation to nutrient availability, predation, and competition. Sommer *et al.* described these patterns as part of the Plankton Ecology Group (PEG) model, with 24 statements constructed from the analysis of numerous systems. The following includes a subset of these statements, as explained by Brönmark and Hansson illustrating succession through a single seasonal cycle:

Winter

1. Increased nutrient and light availability result in rapid phytoplankton growth towards the end of winter. The dominant species, such as diatoms, are small and have quick growth capabilities.

2. These plankton are consumed by zooplankton, which become the dominant plankton taxa.

Spring

3. A clear water phase occurs, as phytoplankton populations become depleted due to increased predation by growing numbers of zooplankton.

Summer

4. Zooplankton abundance declines as a result of decreased phytoplankton prey and increased predation by juvenile fishes.

5. With increased nutrient availability and decreased predation from zooplankton, a diverse phytoplankton community develops.

6. As the summer continues, nutrients become depleted in a predictable order: phosphorus, silica, and then nitrogen. The abundance of various phytoplankton species varies in relation to their biological need for these nutrients.

7. Small-sized zooplankton become the dominant type of zooplankton because they are less vulnerable to fish predation.

Fall

8. Predation by fishes is reduced due to lower temperatures and zooplankton of all sizes increase in number.

Winter

9. Cold temperatures and decreased light availability result in lower rates of primary production and decreased phytoplankton populations. 10. Reproduction in zooplankton decreases due to lower temperatures and less prey.

The PEG model presents an idealized version of this succession pattern, while natural systems are known for their variation.

Latitudinal Patterns

There is a well-documented global pattern that correlates decreasing plant and animal diversity with increasing latitude, that is to say, there are fewer species as one moves towards the poles. The cause of this pattern is one of the greatest puzzles for ecologists today. Theories for its explanation include energy availability, climatic variability, disturbance, competition, etc. Despite this global diversity gradient, this pattern can be weak for freshwater systems compared to global marine and terrestrial systems. This may be related to size, as Hillebrand and Azovsky found that smaller organisms (protozoa and plankton) did not follow the expected trend strongly, while larger species (vertebrates) did. They attributed this to better dispersal ability by smaller organisms, which may result in high distributions globally.

Natural Lake Lifecycles

Lake Creation

Lakes can be formed in a variety of ways, but the most common are discussed briefly below. The oldest and largest systems are the result of tectonic activities. The rift lakes in Africa, for example are the result of seismic activity along the site of separation of two tectonic plates. Ice-formed lakes are created when glaciers recede, leaving behind abnormalities in the landscape shape that are then filled with water. Finally, oxbow lakes are fluvial in origin, resulting when a meandering river bend is pinched off from the main channel.

Natural Extinction

All lakes and ponds receive sediment inputs. Since these systems are not really expanding, it is logical to assume that they will become increasingly shallower in depth, eventually becoming wetlands or terrestrial vegetation. The length of this process should depend upon a combination of depth and sedimentation rate. Moss gives the example of Lake Tanganyika, which reaches a depth of 1500 m and has a sedimentation rate of 0.5 mm/yr. Assuming that sedimentation is not influenced by anthropogenic factors, this system should go extinct in approximately 3 million years. Shallow lentic systems might also fill in as swamps encroach inward from the edges. These processes operate on a much shorter timescale, taking hundreds to thousands of years to complete the extinction process.

Human Impacts

Acidification

Sulfur dioxide and nitrogen oxides are naturally released from volcanoes, organic compounds in the soil, wetlands, and marine systems, but the majority of these compounds come from the combustion of coal, oil, gasoline, and the smelting of ores containing sulfur. These substances dissolve in atmospheric moisture and enter lentic systems as acid rain. Lakes and ponds that contain bedrock that is rich in carbonates have a natural buffer, resulting in no alteration of pH. Systems without this bedrock, however, are very sensitive to acid inputs because they have a low neutralizing capacity, resulting in pH declines even with only small inputs of acid. At a pH of 5–6 algal species diversity and biomass decrease considerably, leading to an increase in water transparency – a characteristic feature of acidified lakes. As the pH continues lower, all fauna becomes less diverse. The most significant feature is the disruption of fish reproduction. Thus, the population is eventually composed of few, old individuals that eventually die and leave the systems without fishes. Acid rain has been especially harmful to lakes in Scandinavia, western Scotland, west Wales and the north eastern United States.

Eutrophication

Eutrophic systems contain a high concentration of phosphorus (~30 μg/L), nitrogen (~1500 μg/L), or both. Phosphorus enters lentic waters from sewage treatment effluents, discharge from raw sewage, or from runoff of farmland. Nitrogen mostly comes from agricultural fertilizers from runoff or leaching and subsequent groundwater flow. This increase in nutrients required for primary producers results in a massive increase of phytoplankton growth, termed a plankton bloom. This bloom decreases water transparency, leading to the loss of submerged plants. The resultant reduction in habitat structure has negative impacts on the species' that utilize it for spawning, maturation and general survival. Additionally, the large number of short-lived phytoplankton result in a massive amount of dead biomass settling into the sediment. Bacteria need large amounts of oxygen to decompose this material, reducing the oxygen concentration of the water. This is especially pronounced in stratified lakes when the thermocline prevents oxygen rich water from the surface to mix with lower levels. Low or anoxic conditions preclude the existence of many taxa that are not physiologically tolerant of these conditions.

Invasive Species

Invasive species have been introduced to lentic systems through both purposeful events (e.g. stocking game and food species) as well as unintentional events (e.g. in ballast water). These organisms can affect natives via competition for prey or habitat, predation, habitat alteration, hybridization, or the introduction of harmful diseases and parasites. With regard to native species, invaders may cause changes in size and age structure, distribution, density, population growth, and may even drive populations to extinction. Examples of prominent invaders of lentic systems include the zebra mussel and sea lamprey in the Great Lakes.

Marine Habitats

The marine environment supplies many kinds of habitats that support marine life. Marine life depends in some way on the saltwater that is in the sea (the term *marine* comes from the Latin *mare*, meaning sea or ocean). A habitat is an ecological or environmental area inhabited by one or more living species.

Marine habitats can be divided into coastal and open ocean habitats. Coastal habitats are found in the area that extends from as far as the tide comes in on the shoreline out to the edge of the continental shelf. Most marine life is found in coastal habitats, even though the shelf area occupies only seven percent of the total ocean area. Open ocean habitats are found in the deep ocean beyond the edge of the continental shelf.

Alternatively, marine habitats can be divided into pelagic and demersal zones. Pelagic habitats are found near the surface or in the open water column, away from the bottom of the ocean. Demersal habitats are near or on the bottom of the ocean. An organism living in a pelagic habitat is said to be a pelagic organism, as in pelagic fish. Similarly, an organism living in a demersal habitat is said to be a demersal organism, as in demersal fish. Pelagic habitats are intrinsically shifting and ephemeral, depending on what ocean currents are doing.

Marine habitats can be modified by their inhabitants. Some marine organisms, like corals, kelp, mangroves and seagrasses, are ecosystem engineers which reshape the marine environment to the point where they create further habitat for other organisms.

Overview

Only 29 percent of the world surface is land. The rest is ocean, home to the marine habitats. The oceans are nearly four kilometres deep on average and are fringed with coastlines that run for nearly 380,000 kilometres.

In contrast to terrestrial habitats, marine habitats are shifting and ephemeral. Swimming organisms find areas by the edge of a continental shelf a good habitat, but only while upwellings bring nutrient rich water to the surface. Shellfish find habitat on sandy beaches, but storms, tides and currents mean their habitat continually reinvents itself.

The presence of seawater is common to all marine habitats. Beyond that many other things determine whether a marine area makes a good habitat and the type of habitat it makes. For example:

- temperature – is affected by geographical latitude, ocean currents, weather, the discharge of rivers, and by the presence of hydrothermal vents or cold seeps

- sunlight – photosynthetic processes depend on how deep and turbid the water is

- nutrients – are transported by ocean currents to different marine habitats from land run-off, or by upwellings from the deep sea, or they sink though the sea as marine snow

- salinity – varies, particularly in estuaries or near river deltas, or by hydrothermal vents.

- dissolved gases – oxygen levels in particular, can be increased by wave actions and decreased during algal blooms.

- acidity – this is partly to do with dissolved gases above, since the acidity of the ocean is largely controlled by how much carbon dioxide is in the water.

- turbulence – ocean waves, fast currents and the agitation of water affect the nature of habitats.

- cover – the availability of cover such as the adjacency of the sea bottom, or the presence of floating objects.

- the occupying organisms themselves – since organisms modify their habitats by the act of occupying them, and some, like corals, kelp, mangroves and seagrasses, create further habitats for other organisms.

Ocean	Area million km²	%	Volume million cu km	%	Mean depth km	Max depth km	Coastline km
Pacific Ocean	155.6	46.4	679.6	49.6	4.37	10.924	135,663
Atlantic Ocean	76.8	22.9	313.4	22.5	4.08	8.605	111,866
Indian Ocean	68.6	20.4	269.3	19.6	3.93	7.258	66,526
Southern Ocean	20.3	6.1	91.5	6.7	4.51	7.235	17,968
Arctic Ocean	14.1	4.2	17.0	1.2	1.21	4.665	45,389
Overall	335.3		1370.8		4.09	10.924	377,412

Land runoff, pouring into the sea, can contain nutrients

The ocean occupies 71 percent of the world surface, averaging nearly four kilometres in depth. There are five major oceans, of which the Pacific Ocean is nearly as large as the rest put together. Coastlines fringe the land for nearly 380,000 kilometres.

Marine habitats can be broadly divided into pelagic and demersal habitats. Pelagic habitats are the habitats of the open water column, away from the bottom of the ocean. Demersal habitats are the habitats that are near or on the bottom of the ocean. An organism living in a pelagic habitat is said to be a pelagic organism, as in pelagic fish. Similarly, an organism living in a demersal habitat is said to be a demersal organism, as in demersal fish. Pelagic habitats are intrinsically ephemeral, depending on what ocean currents are doing.

The land-based ecosystem depends on topsoil and fresh water, while the marine ecosystem depends on dissolved nutrients washed down from the land.

Ocean deoxygenation poses a threat to marine habitats, due to the growth of low oxygen zones.

Ocean Currents

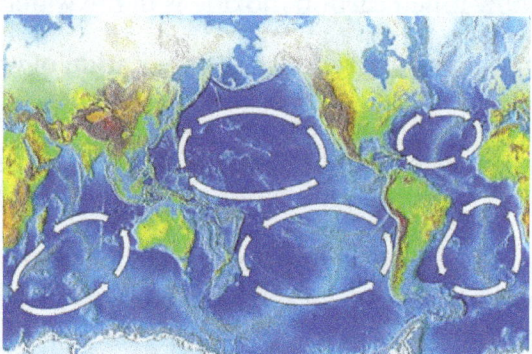

Ocean gyres rotate clockwise in the north and counterclockwise in the south

In marine systems, ocean currents have a key role determining which areas are effective as habitats, since ocean currents transport the basic nutrients needed to support marine life. Plankton are the life forms that inhabit the ocean that are so small (less than 2 mm) that they cannot effectively propel themselves through the water, but must drift instead with the currents. If the current carries the right nutrients, and if it also flows at a suitably shallow depth where there is plenty of sunlight, then such a current itself can become a suitable habitat for photosynthesizing tiny algae called phytoplankton. These tiny plants are the primary producers in the ocean, at the start of the food chain. In turn, as the population of drifting phytoplankton grows, the water becomes a suitable habitat for zooplankton, which feed on the phytoplankton. While phytoplankton are tiny drifting plants, zooplankton are tiny drifting animals, such as the larvae of fish and marine invertebrates. If sufficient zooplankton establish themselves, the current becomes a candidate habitat for the forage fish that feed on them. And then if sufficient forage fish move to the area, it becomes a candidate habitat for larger predatory fish and other marine animals that feed on the forage fish. In this dynamic way, the current itself can, over time, become a moving habitat for multiple types of marine life.

This algae bloom occupies sunlit epipelagic waters off the southern coast of England. The algae are maybe feeding on nutrients from land runoff or upwellings at the edge of the continental shelf

Ocean currents can be generated by differences in the density of the water. How dense water is depends on how saline or warm it is. If water contains differences in salt content or temperature, then the different densities will initiate a current. Water that is saltier or cooler will be denser, and will sink in relation to the surrounding water. Conversely, warmer and less salty water will float to the surface. Atmospheric winds and pressure differences also produces surface currents, waves and seiches. Ocean currents are also generated by the gravitational pull of the sun and moon (tides), and seismic activity (tsunami).

The rotation of the Earth affects the direction ocean currents take, and explains which way the large circular ocean gyres rotate in the image above left. Suppose a current at the equator is heading north. The Earth rotates eastward, so the water possesses that rotational momentum. But the further the water moves north, the slower the earth moves eastward. If the current could get to the North Pole, the earth wouldn't be moving eastward at all. To conserve its rotational momentum, the further the current travels north the faster it must move eastward. So the effect is that the current curves to the right. This is the Coriolis effect. It is weakest at the equator and strongest at the poles. The effect is opposite south of the equator, where currents curve left.

Marine Topography

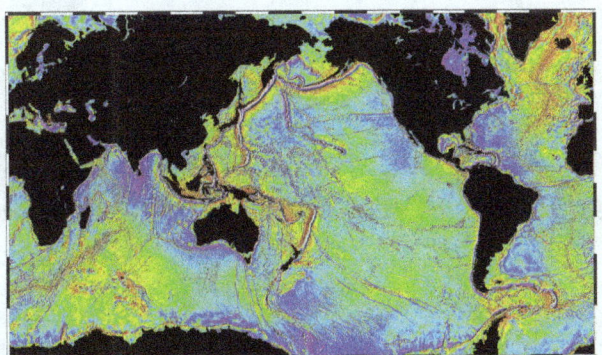

Map of underwater topography (1995 NOAA)

Marine (or seabed or ocean) topography refers to the shape the land has when it interfaces with the ocean. These shapes are obvious along coastlines, but they occur also in significant ways underwater. The effectiveness of marine habitats is partially defined by these shapes, including the way they interact with and shape ocean currents, and the way sunlight diminishes when these landforms occupy increasing depths. Tidal networks depend on the balance between sedimentary processes and hydrodynamics however, anthropogenic influences can impact the natural system more than any physical driver.

Marine topographies include coastal and oceanic landforms ranging from coastal estuaries and shorelines to continental shelves and coral reefs. Further out in the open ocean, they include underwater and deep sea features such as ocean rises and seamounts. The submerged surface has mountainous features, including a globe-spanning mid-ocean ridge system, as well as undersea volcanoes, oceanic trenches, submarine canyons, oceanic plateaus and abyssal plains.

The mass of the oceans is approximately 1.35×10^{18} metric tons, or about 1/4400 of the total mass of the Earth. The oceans cover an area of 3.618×10^8 km² with a mean depth of 3,682 m, resulting in an estimated volume of 1.332×10^9 km³.

Biomass

One measure of the relative importance of different marine habitats is the rate at which they produce biomass.

Producer	Biomass productivity (gC/m²/yr)	Total area (million km²)	Total production (billion tonnes C/yr)	Comment
swamps and marshes	2,500			Includes freshwater
coral reefs	2,000	0.28	0.56	
algal beds	2,000			
river estuaries	1,800			
open ocean	125	311	39	

Coastal

Coastlines can be volatile habitats

Marine coasts are dynamic environments which constantly change, like the ocean which partially shape them. The Earth's natural processes, including weather and sea level change, result in the erosion, accretion and resculpturing of coasts as well as the flooding and creation of continental shelves and drowned river valleys.

The main agents responsible for deposition and erosion along coastlines are waves, tides and currents. The formation of coasts also depends on the nature of the rocks they are made of – the harder the rocks the less likely they are to erode, so variations in rock hardness result in coastlines with different shapes.

Tides often determine the range over which sediment is deposited or eroded. Areas with high tidal ranges allow waves to reach farther up the shore, and areas with lower tidal ranges produce deposition at a smaller elevation interval. The tidal range is influenced by the size and shape of the coastline. Tides do not typically cause erosion by themselves; however, tidal bores can erode as the waves surge up river estuaries from the ocean.

Waves erode coastline as they break on shore releasing their energy; the larger the wave the more energy it releases and the more sediment it moves. Sediment deposited by waves comes from eroded cliff faces and is moved along the coastline by the waves. Sediment deposited by rivers is the dominant influence on the amount of sediment located on a coastline.

Shores that look permanent through the short perceptive of a human lifetime are in fact among the most temporary of all marine structures.

The sedimentologist Francis Shepard classified coasts as *primary* or *secondary*.

- Primary coasts are shaped by non-marine processes, by changes in the land form. If a coast is in much the same condition as it was when sea level was stabilised after the last ice age, it is called a primary coast. "Primary coasts are created by erosion (the wearing away of soil or rock), deposition (the buildup of sediment or sand) or tectonic activity (changes in the structure of the rock and soil because of earthquakes). Many of these coastlines were formed as the sea level rose during the last 18,000 years, submerging river and glacial valleys to form bays and fjords." An example of a primary coast is a river delta, which forms when a river deposits soil and other material as it enters the sea.

- Secondary coasts are produced by marine processes, such as the action of the sea or by creatures that live in it. Secondary coastlines include sea cliffs, barrier islands, mud flats, coral reefs, mangrove swamps and salt marshes.

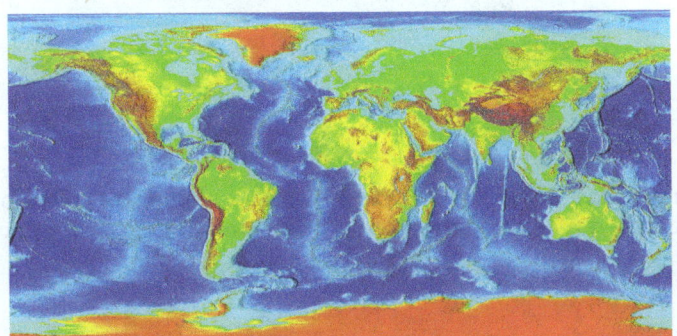

The global continental shelf, highlighted in cyan, defines the extent of coastal habitats, and occupies 5% of the total world area.

Continental coastlines usually have a continental shelf, a shelf of relatively shallow water, less than 200 metres deep, which extends 68 km on average beyond the coast. Worldwide, continental shelves occupy a total area of about 24 million km² (9 million sq mi), 8% of the ocean's total area and nearly 5% of the world's total area. Since the continental shelf is usually less than 200 metres deep, it follows that coastal habitats are generally photic, situated in the sunlit epipelagic zone. This means the conditions for photosynthetic processes so important for primary production, are available to coastal marine habitats. Because land is nearby, there are large discharges of nutrient rich land runoff into coastal waters. Further, periodic upwellings from the deep ocean can provide cool and nutrient rich currents along the edge of the continental shelf.

As a result, coastal marine life is the most abundant in the world. It is found in tidal pools, fjords and estuaries, near sandy shores and rocky coastlines, around coral reefs and on or above the continental shelf. Coastal fish include small forage fish as well as the larger predator fish that feed on them. Forage fish thrive in inshore waters where high productivity results from upwelling and shoreline run off of nutrients. Some are partial residents that spawn in streams, estuaries and bays, but most complete their life cycle in the zone. There can also be a mutualism between species that occupy adjacent marine habitats. For example, fringing reefs just below low tide level have a mutually beneficial relationship with mangrove forests at high tide level and sea grass meadows in between: the reefs protect the mangroves and seagrass from strong currents and waves that would damage them or erode the sediments in which they are rooted, while the mangroves and seagrass protect the coral from large influxes of silt, fresh water and pollutants. This additional level of variety in the environment is beneficial to many types of coral reef animals, which for example may feed in the sea grass and use the reefs for protection or breeding.

Coastal habitats are the most visible marine habitats, but they are not the only important marine habitats. Coastlines run for 380,000 kilometres, and the total volume of the ocean is 1,370 million cu km. This means that for each metre of coast, there is 3.6 cu km of ocean space available somewhere for marine habitats.

Waves and currents shape the intertidal shoreline, eroding the softer rocks and transporting and grading loose particles into shingles, sand or mud

Intertidal

Intertidal zones, those areas close to shore, are constantly being exposed and covered by the ocean's tides. A huge array of life lives within this zone.

Shore habitats range from the upper intertidal zones to the area where land vegetation takes prominence. It can be underwater anywhere from daily to very infrequently. Many species here are scavengers, living off of sea life that is washed up on the shore. Many land animals also make much use of the shore and intertidal habitats. A subgroup of organisms in this habitat bores and grinds exposed rock through the process of bioerosion.

Sandy Shores

Sandy shores provide shifting homes to many species

Sandy shores, also called beaches, are coastal shorelines where sand accumulates. Waves and currents shift the sand, continually building and eroding the shoreline. Longshore currents flow parallel to the beaches, making waves break obliquely on the sand. These currents transport large amounts of sand along coasts, forming spits, barrier islands and tombolos. Longshore currents also commonly create offshore bars, which give beaches some stability by reducing erosion.

Sandy shores are full of life, The grains of sand host diatoms, bacteria and other microscopic creatures. Some fish and turtles return to certain beaches and spawn eggs in the sand. Birds habitat beaches, like gulls, loons, sandpipers, terns and pelicans. Aquatic mammals, such sea lions, recuperate on them. Clams, periwinkles, crabs, shrimp, starfish and sea urchins are found on most beaches.

Sand is a sediment made from small grains or particles with diameters between about 60 µm and 2 mm. Mud is a sediment made from particles finer than sand. This small particle size means that mud particles tend to stick together, whereas sand particles do not. Mud is not easily shifted by waves and currents, and when it dries out, cakes into a solid. By contrast, sand is easily shifted by waves and currents, and when sand dries out it can be blown in the wind, accumulating into shifting sand dunes. Beyond the high tide mark, if the beach is low-lying, the wind can form rolling hills of sand dunes. Small dunes shift and reshape under the influence of the wind while larger dunes stabilise the sand with vegetation.

Ocean processes grade loose sediments to particle sizes other than sand, such as gravel or cobbles. Waves breaking on a beach can leave a berm, which is a raised ridge of coarser pebbles or sand, at the high tide mark. Shingle beaches are made of particles larger than sand, such as cobbles, or small stones. These beaches make poor habitats. Little life survives because the stones are churned and pounded together by waves and currents.

Rocky Shores

Tidepools on rocky shores make turbulent habitats for many forms of marine life

The relative solidity of rocky shores seems to give them a permanence compared to the shifting nature of sandy shores. This apparent stability is not real over even quite short geological time scales, but it is real enough over the short life of an organism. In contrast to sandy shores, plants and animals can anchor themselves to the rocks.

Competition can develop for the rocky spaces. For example, barnacles can compete successfully on open intertidal rock faces to the point where the rock surface is covered with them. Barnacles resist desiccation and grip well to exposed rock faces. However, in the crevices of the same rocks, the inhabitants are different. Here mussels can be the successful species, secured to the rock with their byssal threads.

Rocky and sandy coasts are vulnerable because humans find them attractive and want to live near them. An increasing proportion of the humans live by the coast, putting pressure on coastal habitats.

Mudflats

Mudflats become temporary habitats for migrating birds

Mudflats are coastal wetlands that form when mud is deposited by tides or rivers. They are found in sheltered areas such as bays, bayous, lagoons, and estuaries. Mudflats may be viewed geologically as exposed layers of bay mud, resulting from deposition of estuarine silts, clays and marine animal detritus. Most of the sediment within a mudflat is within the intertidal zone, and thus the flat is submerged and exposed approximately twice daily.

Mudflats are typically important regions for wildlife, supporting a large population, although levels of biodiversity are not particularly high. They are of particular importance to migratory birds. In the United Kingdom mudflats have been classified as a Biodiversity Action Plan priority habitat.

Mangrove and Salt Marshes

Mangroves provide nurseries for fish

Mangrove swamps and salt marshes form important coastal habitats in tropical and temperate areas respectively.

Mangroves are species of shrubs and medium size trees that grow in saline coastal sediment habitats in the tropics and subtropics – mainly between latitudes 25° N and 25° S. The saline conditions tolerated by various species range from brackish water, through pure seawater (30 to 40 ppt), to water concentrated by evaporation to over twice the salinity of ocean seawater (up to 90 ppt). There are many mangrove species, not all closely related. The term "mangrove" is used generally to cover all of these species, and it can be used narrowly to cover just mangrove trees of the genus *Rhizophora*.

Mangroves form a distinct characteristic saline woodland or shrubland habitat, called a *mangrove swamp* or *mangrove forest'*. Mangrove swamps are found in depositional coastal environments, where fine sediments (often with high organic content) collect in areas protected from high-energy wave action. Mangroves dominate three quarters of tropical coastlines.

Estuaries

Estuaries occur when rivers flow into a coastal bay or inlet. They are nutrient rich and have a transition zone which moves from freshwater to saltwater.

An estuary is a partly enclosed coastal body of water with one or more rivers or streams flowing into it, and with a free connection to the open sea. Estuaries form a transition zone between river environments and ocean environments and are subject to both marine influences, such as tides, waves, and the influx of saline water; and riverine influences, such as flows of fresh water and sediment. The inflow of both seawater and freshwater provide high levels of nutrients in both the water column and sediment, making estuaries among the most productive natural habitats in the world.

Most estuaries were formed by the flooding of river-eroded or glacially scoured valleys when sea level began to rise about 10,000-12,000 years ago. They are amongst the most heavily populated areas throughout the world, with about 60% of the world's population living along estuaries and the coast. As a result, estuaries are suffering degradation by many factors, including sedimentation from soil erosion from deforestation; overgrazing and other poor farming practices; overfishing; drainage and filling of wetlands; eutrophication due to excessive nutrients from sewage and animal wastes; pollutants including heavy metals, PCBs, radionuclides and hydrocarbons from sewage inputs; and diking or damming for flood control or water diversion.

Estuaries provide habitats for a large number of organisms and support very high productivity. Estuaries provide habitats for salmon and sea trout nurseries, as well as migratory bird populations.

Two of the main characteristics of estuarine life are the variability in salinity and sedimentation. Many species of fish and invertebrates have various methods to control or conform to the shifts in salt concentrations and are termed osmoconformers and osmoregulators. Many animals also burrow to avoid predation and to live in the more stable sedimental environment. However, large numbers of bacteria are found within the sediment which have a very high oxygen demand. This reduces the levels of oxygen within the sediment often resulting in partially anoxic conditions, which can be further exacerbated by limited water flux. Phytoplankton are key primary producers in estuaries. They move with the water bodies and can be flushed in and out with the tides. Their productivity is largely dependent on the turbidity of the water. The main phytoplankton present are diatoms and dinoflagellates which are abundant in the sediment.

Kelp Forests

Kelp forests provide habitat for many marine organisms

Kelp forests are underwater areas with a high density of kelp. They form some of the most productive and dynamic ecosystems on Earth. Smaller areas of anchored kelp are called *kelp beds*. Kelp forests occur worldwide throughout temperate and polar coastal oceans.

Kelp forests provide a unique three-dimensional habitat for marine organisms and are a source for understanding many ecological processes. Over the last century, they have been the focus of extensive research, particularly in trophic ecology, and continue to provoke important ideas that are relevant beyond this unique ecosystem. For example, kelp forests can influence coastal oceanographic patterns and provide many ecosystem services.

However, humans have contributed to kelp forest degradation. Of particular concern are the effects of overfishing nearshore ecosystems, which can release herbivores from their normal population regulation and result in the over-grazing of kelp and other algae. This can rapidly result in transitions to barren landscapes where relatively few species persist.

Frequently considered an ecosystem engineer, kelp provides a physical substrate and habitat for kelp forest communities. In algae (Kingdom: Protista), the body of an individual organism is known as a thallus rather than as a plant (Kingdom: Plantae). The morphological structure of a kelp thallus is defined by three basic structural units:

- The holdfast is a root-like mass that anchors the thallus to the sea floor, though unlike true roots it is not responsible for absorbing and delivering nutrients to the rest of the thallus;

- The stipe is analogous to a plant stalk, extending vertically from the holdfast and providing a support framework for other morphological features;

- The fronds are leaf- or blade-like attachments extending from the stipe, sometimes along its full length, and are the sites of nutrient uptake and photosynthetic activity.

In addition, many kelp species have pneumatocysts, or gas-filled bladders, usually located at the base of fronds near the stipe. These structures provide the necessary buoyancy for kelp to maintain an upright position in the water column.

The environmental factors necessary for kelp to survive include hard substrate (usually rock), high nutrients (e.g., nitrogen, phosphorus), and light (minimum annual irradiance dose > 50 E m^{-2}). Especially productive kelp forests tend to be associated with areas of significant oceanographic upwelling, a process that delivers cool nutrient-rich water from depth to the ocean's mixed surface layer. Water flow and turbulence facilitate nutrient assimilation across kelp fronds throughout the water column. Water clarity affects the depth to which sufficient light can be transmitted. In ideal conditions, giant kelp (*Macrocystis spp.*) can grow as much as 30-60 centimetres vertically per day. Some species such as *Nereocystis* are annual while others like *Eisenia* are perennial, living for more than 20 years. In perennial kelp forests, maximum growth rates occur during upwelling months (typically spring and summer) and die-backs correspond to reduced nutrient availability, shorter photoperiods and increased storm frequency.

Seagrass Meadows

White-spotted puffers like living in seagrass areas

Seagrasses are flowering plants from one of four plant families which grow in marine environments. They are called *seagrasses* because the leaves are long and narrow and are very often green, and because the plants often grow in large meadows which look like grassland. Since seagrasses photosynthesize and are submerged, they must grow submerged in the photic zone, where there is enough sunlight. For this reason, most occur in shallow and sheltered coastal waters anchored in sand or mud bottoms.

Seagrasses form extensive beds or meadows, which can be either monospecific (made up of one species) or multispecific (where more than one species co-exist). Seagrass beds make highly diverse and productive ecosystems. They are home to phyla such as juvenile and adult fish, epiphytic and free-living macroalgae and microalgae, mollusks, bristle worms, and nematodes. Few species were originally considered to feed directly on seagrass leaves (partly because of their low nutritional content), but scientific reviews and improved working methods have shown that seagrass herbivory is a highly important link in the food chain, with hundreds of species feeding on seagrasses worldwide, including green turtles, dugongs, manatees, fish, geese, swans, sea urchins and crabs.

Seagrasses are ecosystem engineers in the sense that they partly create their own habitat. The leaves slow down water-currents increasing sedimentation, and the seagrass roots and rhizomes stabilize the seabed. Their importance to associated species is mainly due to provision of shelter (through their three-dimensional structure in the water column), and due to their extraordinarily high rate of primary production. As a result, seagrasses provide coastal zones with ecosystem services, such as fishing grounds, wave protection, oxygen production and protection against coastal erosion. Seagrass meadows account for 15% of the ocean's total carbon storage.

Coral Reefs

Reefs comprise some of the densest and most diverse habitats in the world. The best-known types of reefs are tropical coral reefs which exist in most tropical waters; however, reefs can also exist in cold water. Reefs are built up by corals and other calcium-depositing animals, usually on top of a rocky outcrop on the ocean floor. Reefs can also grow on other surfaces, which has made it possible to create artificial reefs. Coral reefs also support a huge community of life, including the corals themselves, their symbiotic zooxanthellae, tropical fish and many other organisms.

Much attention in marine biology is focused on coral reefs and the El Niño weather phenomenon. In 1998, coral reefs experienced the most severe mass bleaching events on record, when vast expanses of reefs across the world died because sea surface temperatures rose well above normal. Some reefs are recovering, but scientists say that between 50% and 70% of the world's coral reefs are now endangered and predict that global warming could exacerbate this trend.

Open Ocean

The open ocean is relatively unproductive because of a lack of nutrients, yet because it is so vast, it has more overall primary production than any other marine habitat. Only about 10 percent of marine species live in the open ocean. But among them are the largest and fastest of all marine animals, as well as the animals that dive the deepest and migrate the longest. In the depths lurk animal that, to our eyes, appear hugely alien.

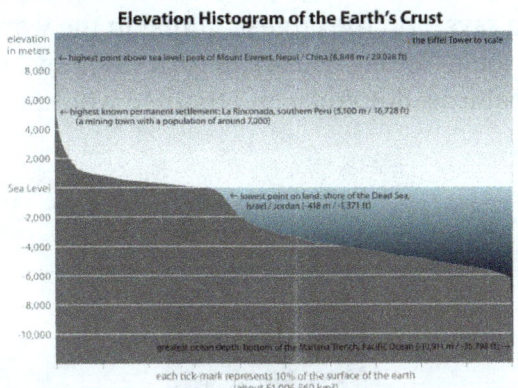

Elevation-area graph showing the proportion of land area at given heights
and the proportion of ocean area at given depths

Surface Waters

In the open ocean, sunlit surface epipelagic waters get enough light for photosynthesis, but there are often not enough nutrients. As a result, large areas contain little life apart from migrating animals.

The surface waters are sunlit. The waters down to about 200 metres are said to be in the epipelagic zone. Enough sunlight enters the epipelagic zone to allow photosynthesis by phytoplankton. The epipelagic zone is usually low in nutrients. This partially because the organic debris produced in the zone, such as excrement and dead animals, sink to the depths and are lost to the upper zone. Photosynthesis can happen only if both sunlight and nutrients are present.

In some places, like at the edge of continental shelves, nutrients can upwell from the ocean depth, or land runoff can be distributed by storms and ocean currents. In these areas, given that both sunlight and nutrients are now present, phytoplankton can rapidly establish itself, multiplying so fast that the water turns green from the chlorophyll, resulting in an algal bloom. These nutrient rich surface waters are among the most biologically productive in the world, supporting billions of tonnes of biomass.

"Phytoplankton are eaten by zooplankton - small animals which, like phytoplankton, drift in the ocean currents. The most abundant zooplankton species are copepods and krill: tiny crustaceans that are the most numerous animals on Earth. Other types of zooplankton include jelly fish and the larvae of fish, marine worms, starfish, and other marine organisms". In turn, the zooplankton are eaten by filter-feeding animals, including some seabirds, small forage fish like herrings and sardines, whale sharks, manta rays, and the largest animal in the world, the blue whale. Yet again, moving up the foodchain, the small forage fish are in turn eaten by larger predators, such as tuna, marlin, sharks, large squid, seabirds, dolphins, and toothed whales.

Deep Sea

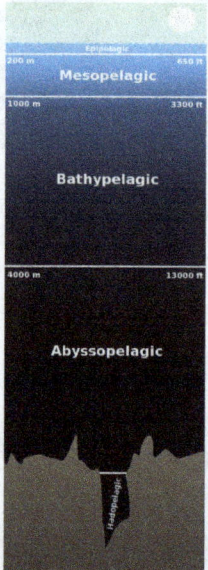

Scale diagram of the layers of the pelagic zone

The deep sea starts at the aphotic zone, the point where sunlight loses most of its energy in the water. Many life forms that live at these depths have the ability to create their own light a unique evolution known as bio-luminescence.

In the deep ocean, the waters extend far below the epipelagic zone, and support very different types of pelagic life forms adapted to living in these deeper zones.

Much of the aphotic zone's energy is supplied by the open ocean in the form of detritus. In deep water, marine snow is a continuous shower of mostly organic detritus falling from the upper layers of the water column. Its origin lies in activities within the productive photic zone. Marine snow includes dead or dying plankton, protists (diatoms), fecal matter, sand, soot and other inorganic dust. The "snowflakes" grow over time and may reach several centimetres in diameter, travelling for weeks before reaching the ocean floor. However, most organic components of marine snow are consumed by microbes, zooplankton and other filter-feeding animals within the first 1,000 metres of their journey, that is, within the epipelagic zone. In this way marine snow may be considered the foundation of deep-sea mesopelagic and benthic ecosystems: As sunlight cannot reach them, deep-sea organisms rely heavily on marine snow as an energy source.

Some deep-sea pelagic groups, such as the lanternfish, ridgehead, marine hatchetfish, and lightfish families are sometimes termed *pseudoceanic* because, rather than having an even distribution in open water, they occur in significantly higher abundances around structural oases, notably seamounts and over continental slopes. The phenomenon is explained by the likewise abundance of prey species which are also attracted to the structures.

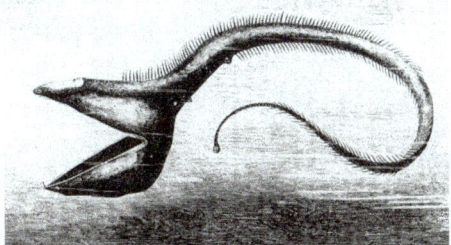

The umbrella mouth gulper eel can swallow a fish much larger than itself

The fish in the different pelagic and deep water benthic zones are physically structured, and behave in ways, that differ markedly from each other. Groups of coexisting species within each zone all seem to operate in similar ways, such as the small mesopelagic vertically migrating plankton-feeders, the bathypelagic anglerfishes, and the deep water benthic rattails.

Ray finned species, with spiny fins, are rare among deep sea fishes, which suggests that deep sea fish are ancient and so well adapted to their environment that invasions by more modern fishes have been unsuccessful. The few ray fins that do exist are mainly in the Beryciformes and Lampriformes, which are also ancient forms. Most deep sea pelagic fishes belong to their own orders, suggesting a long evolution in deep sea environments. In contrast, deep water benthic species, are in orders that include many related shallow water fishes.

The umbrella mouth gulper is a deep sea eel with an enormous loosely hinged mouth. It can open its mouth wide enough to swallow a fish much larger than itself, and then expand its stomach to accommodate its catch.

Sea Floor

Vents and Seeps

Zooarium chimney provides a habitat for vent biota

Hydrothermal vents along the mid-ocean ridge spreading centers act as oases, as do their opposites, cold seeps. Such places support unique marine biomes and many new marine microorganisms and other lifeforms have been discovered at these locations.

Trenches

The deepest recorded oceanic trenches measure to date is the Mariana Trench, near the Philippines, in the Pacific Ocean at 10,924 m (35,838 ft). At such depths, water pressure is extreme and there is no sunlight, but some life still exists. A white flatfish, a shrimp and a jellyfish were seen by the American crew of the bathyscaphe *Trieste* when it dove to the bottom in 1960.

Seamounts

Marine life also flourishes around seamounts that rise from the depths, where fish and other sea life congregate to spawn and feed.

Nursery Habitat

In marine environments, a nursery habitat is a subset of all habitats where juveniles of a species occur, having a greater level of productivity per unit area than other juvenile habitats (Beck et al. 2001). Mangroves, salt marshes and seagrass are typical nursery habitats for a range of marine species. Some species will use nonvegetated sites, such as Yellow Eyed Mullet, Blue Sprat and Flounder.

Overview

The nursery habitat hypothesis states that the contribution per unit area of a nursery habitat is greater than for other habitats used by juveniles for the species. Productivity may be measured by density, survival, growth and movement to adult habitat (Beck et al. 2001).

There are two general models for the location of juvenile habitats within the total range for a species which reflect life history strategies of the species. These are the Classic Concept: Juveniles and Adults in separate habitats. Juveniles migrate to adult habitat. General Concept: overlap of juvenile and adult habitats.

Some marine species do not have juvenile habitats, e.g. arthropods and scallops. Commonly fish, ells, some lobsters, blue crabs (and so forth) do have distinct juvenile habitats, whether with or without overlap with adult habitats.

In terms of management, use of the nursery role hypothesis may be limiting as it excludes some potentially important nursery sites. In these cases the Effective Juvenile Habitat concept may be more useful. This defines a nursery as that which supplies a higher percentage of individuals to adult populations.

Identification and subsequent management of nursery habitats may be important in supporting off shore fisheries and ensuring species survival into the future. If we are unable to preserve nursery habitats, recruitment of juveniles into adult populations may decline, reducing population numbers and compromising the survival of species for biodiversity and human harvesting.

Determination

In order to determine the nursery habitat for a species, all habitats used by juveniles must be surveyed. This may include kelp forest, seagrass, mangroves, tidal flat, mudflat, wetland, salt marsh and oyster reef. While density may be an indicator of productivity, it is suggested that alone, density does not adequately provide evidence of the role of a habitat as a nursery. Recruitment biomass from juvenile to adult population is the best measure of movement between the two habitats.

Consider also biotic, abiotic and landscape variability in the value of nursery habitats. This may be an important consideration when looking at which sites to manage and protect. Biotic factors include: structural complexity, food availability, larval settlement cues, competition, and predation. Abiotic: temperature, salinity, depth, dissolved oxygen, freshwater inflow, retention zone and disturbance. Landscape factors involve: proximity of juvenile and adult habitats, access to larvae, number of adjacent habitats, patch shape, area and fragmentation. The effects of these factors may be positive or negative depending on species and broader environmental conditions at any given time.

It may be more holistic to consider temporal variation in habitats used as nurseries, and incorporating temporal scales into any testing is important. Also consider assemblages of species. Single species approaches may not be able to be used to adequately manage systems appropriately.

Acosta and Butler conducted experimental observation of spiny lobster to determine which habitats are used as nurseries. Mangroves are used as preferred nursery habitat when coral density is low. Predation on newly settled larvae was lower in mangrove than in seagrass beds and coral crevices. In comparison, Pipefish prefer seagrass over algae and sand habitats. King George Whiting have a more complex pattern of development. Settlement is preferred in seagrass and algae. Growth stages are primarily preferred in reef algae. 4 months post settlement, they move into unvegetated habitats (Jenkins and Wheatley, 1998).

Classification of Habitats

In studies of the ecology of freshwater rivers, habitats are classified as upland and lowland. Upland habitats are cold, clear, rocky, fast flowing rivers in mountainous areas; lowland habitats are warm, slow flowing rivers found in relatively flat lowland areas, with water that is frequently coloured by sediment and organic matter.

Classifying rivers and streams as upland or lowland is important in freshwater ecology as the two types of river habitat are very different, and usually support very different populations of fish and invertebrate species.

Upland

In freshwater ecology, upland rivers and streams are the fast flowing rivers and streams that drain elevated or mountainous country, often onto broad alluvial plains (where they become lowland rivers). However, altitude is not the sole determinant of whether a river is upland or lowland. Arguably the most important determinants are that of stream power and course gradient. Rivers with a course that drops in altitude rapidly will have faster water flow and higher stream power or "force of water". This in turn produces the other characteristics of an upland river - an incised course, a river bed dominated by bedrock and coarse sediments, a riffle and pool structure and cooler water temperatures.

Upland lake

Rivers with a course that drops in altitude very slowly will have slower water flow and lower force. This in turn produces the other characteristics of a lowland river - a meandering course lacking rapids, a river bed dominated by fine sediments and higher water temperatures. Lowland rivers tend to carry more suspended sediment and organic matter as well, but some lowland rivers have periods of high water clarity in seasonal low flow periods.

The generally clear, cool, fast-flowing waters and bedrock and coarse sediment beds of upland rivers encourage fish species with limited temperature tolerances, high oxygen needs, strong swimming ability and specialized reproductive strategies to prevent eggs or larvae being swept away. These characteristics also encourage invertebrate species with limited temperature tolerances, high oxygen needs and ecologies revolving around coarse sediments and interstices or "gaps" between those coarse sediments.

Lowland

The generally more turbid, warm, slow-flowing waters and fine sediment beds of lowland rivers encourage fish species with broad temperature tolerances and greater tolerances to low oxygen levels, and life history and breeding strategies adapted to these and other traits of lowland rivers. These characteristics also encourage invertebrate species with broad temperature tolerances and greater tolerances to low oxygen levels and ecologies revolving around fine sediments or alternative habitats such as submerged woody debris ("snags") or submergent macrophytes ("water weed").

Lowland river

There are four main constituents of the living environment that form the freshwater ecosystem, they are as follows.

- Elements and Compounds of the ecosystem that are absorbed by organisms that are required as a food source or for respiration. Many of these compounds are required by plants and passed along the food chain.

- Plants which are autotrophic by nature, meaning that they synthesize food by harnessing energy from inorganic compounds (plants do so by photosynthesis and the sun); this is done via photosynthesis. These plants (and some bacteria) are the primary producers, as they produce (and introduce) new energy into the ecosystem.

- Consumers, which are the organisms that feed on other organisms as a source of food. These may be primary consumers who feed from the plant material or secondary consumers who feed on the primary consumers.

- Decomposers attain their energy by breaking down dead organic material (detritus), and during this reaction, release critical elements and compounds which in turn are required by plants.

Biotic and Abiotic Factors – Freshwater Ecology

Abiotic factors are essentially non-living components that affect the living organisms of the freshwater community.

When an ecosystem is barren and unoccupied, new organisms colonizing the environment rely on favorable environmental conditions in the area to allow them to successfully live and reproduce. These environmental factors are abiotic factors. When a variety of species are present in such an ecosystem, the consequent actions of these species can affect the lives of fellow species in the area; these factors are deemed biotic factors.

The light from the sun is a major constituent of a freshwater ecosystem, providing light for the primary producers, plants. There are many factors which can affect the intensity and length of time that the ecosystem is exposed to sunlight;

- Aspect - The angle of incidence at which light strikes the surface of the water. During the day when the sun is high in the sky, more light can be absorbed into the water due to the directness of the light. At sunset, light strikes the water surface more acutely, and less water is absorbed. The aspect of the sun during times of the day will vary depending on the time of the year.

- Cloud Cover - The cloud cover of an area will inevitably affect intensity and length of time that light strikes the water of a freshwater ecosystem. Species of plants rely on a critical period of time where they receive light for photosynthesis.

- Season - The 4 seasons in an ecosystem are very different, and this is because less light and heat is available from the sun in Winter and vice versa for Summer, therefore these varying conditions will affect which organisms are suited to them.

- Location - The extreme latitudes receive 6 months of sunlight and 6 months of darkness, while the equator receives roughly 12 hours of sunlight and darkness each day. This sort of variance greatly affects what type of organisms would occupy freshwater ecosystems due to these differences.

- Altitude - For every one thousand metres above sea level, average temperature drops by one degree Celsius. Altitude will also affect the aspect of which sunlight hits the freshwater ecosystem, therefore playing a part on which organisms will occupy it.

Many abiotic factors can play a part in determining the end product, which organisms live and succeed in the freshwater ecosystem. The sun provides light for photosynthesis, but also provides heat giving a suitable temperature for organisms to thrive in. The temperature of a freshwater environment can directly affect the environment as a whole and the organisms that occupy it.

Enzymes operate best at an optimum temperature, and any deviation from this temperature 'norm' will result in below optimum respiration in the organism. All aquatic life are ectotherms, meaning their body temperature varies directly with its environments.

Temperature affects the density of substances, and changes in the density of water means more or less resistance for animals who are travelling in the freshwater environment.

Abiotic Factors - Water Conditions

Evidently, the light and heat from the sun play an important role in providing suitable conditions. However, the water conditions also inevitably have an effect on life in the ecosystem. A still body of water will inevitably be disturbed by various factors, which will affect the distribution of organisms in the water. Wind is considered to be the prime factor responsible for disturbing water, though changes in temperature can create convection currents where temperature is evened out across the body of water via this movement.

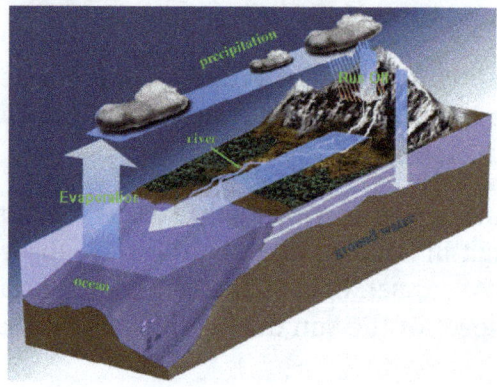

Naturally, a river will have water movement as water succumbs to gravity and moves downstream. These are relatively constant factors that affect water movement though, for example, human intervention can also cause water movement. The surface tension of the water will also affect the organisms that occupy the area, depending on the cohesion of water at the surface; it can affect the amount of oxygen that reaches organisms living below the water surface.

These factors all affect the way of life for organisms occupying such a freshwater ecosystem. On a more molecular level, the chemical compositions of the water, soil and surrounding air also play a part in determining the face of the ecosystem.

The oxygen concentration of the water and the surrounding air will have great bearing on which organisms can survive in a particular environment. Oxygen is required for aerobic respiration in animals, and the concentration of oxygen in an area is determined by many factors, including temperature and abundance of organisms for example.

Many chemical reactions and cellular processes rely on the availability of oxygen; therefore the concentration of oxygen in the ecosystem will inevitably alter the ecosystem itself. The same applies to carbon dioxide concentration. CO_2 is required for photosynthesis, and can also affect the pH of the water for example.

The study of ecology in freshwater is usually divided into 2 categories, lentic (still) and lotic (running) water. These two bodies of water also have a bearing on which organisms are likely to occupy the area.

Freshwater Communities & Lentic Waters

Zones of a Lake

Lentic (still water) communities can vary greatly in appearance; anything from a small temporary puddle to a large lake is capable of supporting life to some extent.

The creation of many of today's long standing freshwater lentic environments are a result of geological changes over a long period of time, notably glacial movement, erosion, volcanic activity, and to an extent, human intervention.

The consequence of these actions results in troughs in the landscape where water can accumulate and be sustained over time. The size and depth of a still body of water are major factors in determining the characteristics of that ecosystem, and will continually be altered by some of the causes mentioned above over a long period of time.

One of the important elements of a still water environment is the overall effect that temperature has on it. The heat from the sun takes longer to heat up a body of water as opposed to heating up dry land. This means that temperature changes in the water are more gradual, particularly so in more vast areas of water. When this freshwater ecosystem is habitable, many factors will come into play determining the overall make up of the environment which organisms will have to adapt to.

As with osmosis, temperature will even out across a particular substance over time, and this applies to a still body of water. Sunlight striking the water will heat up the surface, and over time will create a temperature difference between the surface and basin in the body of water. This temperature difference will vary depending on the overall surface area of the water and its depth.

Over time, two distinctly different layers of water become established, separated by a large temperature difference and providing unique ecological niches for organisms. This process is called stratification, where the difference in temperature between surface and water bed are so different they can easily be distinguished apart. The surface area is deemed the epilimnion, which is warmed water as a result of direct contact with sunlight. The lower layer is deemed the hypolimnion, found below the water surface, and due to increased depth, receives less heat from the sun and therefore results in the colder water underneath.

Some factors can affect the amount of light received by autotrophic organisms (organisms that perform photosynthesis) can affect their level of photosynthesis and respiration, hence affect their abundance and therefore and subsequent species that rely on them.

Organic material and sediment can enter the still water environment via dead organisms in the area, and water flowing into the area from hills and streams. Buoyant material will also block out light required by the primary producers of the ecosystem.

When water moves, the friction caused by the moving water against the water bed and its banks will result in disturbing loose sediment. Depending on the weight of this sediment, heavier particles will slowly sink back to the bottom of the body of water while lighter materials will remain suspended in the water. The lightest material will rise to the surface, resulting in less light available to organisms underneath the surface.

Naturally, the consequences of the above will result in less light for organisms that rely on photosynthesis as a means of food, and subsequently means that organisms that feed on these autotrophic organisms will soon find that their food source is less freely available.

Another major factor affecting still water communities is the oxygen concentration of the surrounding area. Oxygen concentration is primarily affected by three factors

- The surface area which the body of water is exposed to the open air environment

- The circulation of water, chiefly due to temperature differentiations in different areas of the water body (convection currents)

- Oxygen created as a result of respiration, consumption, and the oxygen consumed by animals and bacteria.

Temperature can also affect the concentration of oxygen available, which in turn, means that the depth of the water will therefore also have an effect. In turn, carbon dioxide levels, which are closely related to the oxygen levels available, will be required by organisms undergoing photosynthesis. The availability of these will affect the organisms in the ecosystem. Their relationships with temperature will also affect their availability. Evidently, some of these factors vary through different conditions, and changes in one of the factors usually results in changes with the others. This is also the case of pH, for example, as an increase in carbon dioxide results in a drop of pH.

Still Water Animals

Through millions of years of evolution, animals living in an aquatic environment have diversified to occupy the ecological niches available in the ecosystem. When studying the habitats of these particular organisms, three main areas of the freshwater environment can be distinctly classified.

- The Profundal Region - An area of still water that receives no sunlight therefore lacks autotrophic creatures. The animals in this zone rely on organic material as a means of food, which is sourced from the more energy rich areas above the profundal region.

- The Pelagic Region - The pelagic region can be found below the surface water, and is defined by the light that is available to it. The pelagic region does not include areas near the shore or sea bed.

- The Benthic Region - The benthic region incorporates all the freshwater environment in contact with land, barring the shallow shore areas. The benthic region is capable of hosting a large volume of organisms, as nutrient and mineral rich sediments are available as a food source while part of the benthic region can occupy the euphotic zone, the area of water where light is available. This will allow an ecological niche for autotrophic organisms which in turn can be a food source for herbivores.

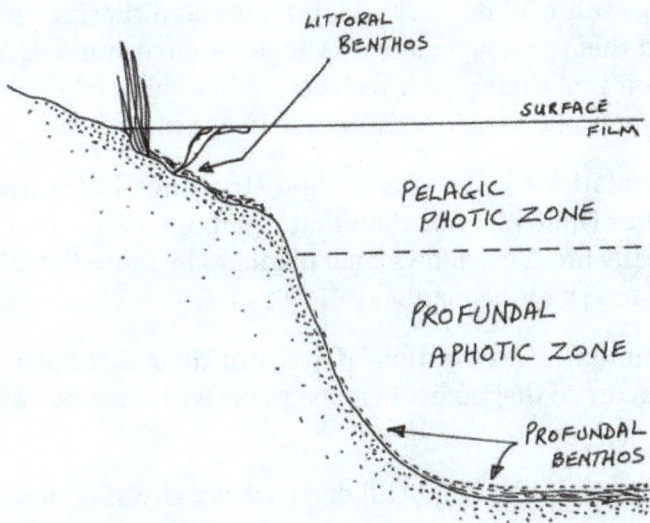

Another distinctive niche for the animal community is that above (epineuston) and below (hyponeuston) the water surface. Epineustic animals receive food from the surrounding hydrosere vegetation, where small animals fall into the water from vegetation and are preyed upon by these epineustic animals.

Below these surface dwelling animals are a collective of animals called the nekton, which live in the pelagic and profundal regions, though rise to the pelagic regions to feed upon these epineustic animals. Fish are included in this nekton community, which play a vital cog in these freshwater communities. Some of these fish are only temporary members of the community, as they move between fresh and salt water. Anadromous fish spawn in freshwater, but live much of their lives in salt water. Catadromous fish are the opposite of this, and spend much of their lives in the freshwater community. Each way, the fish present in the environment at any time form the link between the upper and lower layers of the freshwater community.

Freshwater Lentic Communities & Animals

Plants that live partially or completely submerged in water are deemed hydrophytes. A form of symbiosis occurs with these hydrophyte plants, which provide means for algae and other organisms to survive in the surrounding environment. This is because the hydrophytes provide the conditions for the likes of algae and bacteria to survive in the environment. In return, herbivore animals tend to feed on this rich blanket of algae as opposed to the plants themselves, therefore protecting them from being consumed.

Animals in this environment feed on these algae, and also upon the detritus matter, the organic material that is rich on the water bed. It is an area of abundant organic material because the plants that survive in this area provide a source of food, and also a source of shelter which can provide protection from predators or a location to hatch offspring in a closed protected area.

The ecological niche alongside the still water banks is occupied by plants called hydroseres, which are partially or totally submerged by water along the banks. Some of these hydroseres are rooted in the water, though some of their leaves penetrate the water surface, while others float on the surface, one side in contact with the water, the other side in contact with the open air environment. In essence, hydroseres possess evolutionary adaptations and dithering respiration rates from land plants that have allowed them to adapt in live in such an environment. Such evolutionary adaptation in plants has meant that their physical structure has changed to suit the environment, and therefore making freshwater plants distinctly unique in appearance.

An example of these adaptations is the lack of rigid structures in freshwater plants. This is due to the density of the water (much higher than that of an open air environment), which 'pushes' against the plant in its daily life. This allows such plants to be more flexible against oncoming water tides, and prevents damage to the plant.

As plants require a minimum concentration of gases in their diet such as carbon dioxide, they require a degree of buoyancy so that contact can be made with the open air environment. Adaptations may include;

- Air Spaces - Air spaces in the plant will decrease density and increase buoyancy.

- Broad Leaves - Broader leaves will spread their weight more evenly across the water surface allowing them to float.

- Waxy Cuticle - On the upper half to allow water to run off the surface to prevent the weight of the water dragging the leaves under the surface.

In still water plants, the method of transpiration as a whole is altered in freshwater plants, due to the abundance of water in their external environment, or in the case of some, uptake of water from a wet environment, but loss of water via their leaves in the open air environment.

An example of transpiration problems for such plants is as follows;

- The plant lives in a marshy environment, where roots uptake water from soaked ground, allowing plenty of water to be up taken and transported up and across the plant.

- The difference in water concentration between the plants' leaves and the open air environment is so great that much of the water absorbed is lost to the external environment, meaning the plant loses water rapidly.

- Such a problem is solved by evolutionary adaptations. These adaptations essentially address the issue of re-balancing the critical deviations between the water that is absorbed and lost in a plant.

Freshwater Plants & Nutrients

On top of the need for plants to maintain a suitable water concentration in plant cells, they also require various nutrients which are found in the nutrient rich soil and the surrounding waters. In addition to the carbon, hydrogen and oxygen required for photosynthesis, plants require a range of macro-elements, notably magnesium (Mg), nitrogen (N), phosphorous (P) and potassium (K). Some of these elements, notably the gases, are readily available in the atmosphere, while carbon dioxide is produced from decomposing organic matter. Other elements are readily available in the soil, with nutrients becoming available from decomposing matter adding to the fertility of the surrounding soil. Oxygen becomes available from the photosynthetic activities of plants, which provide the link between oxygen and carbon dioxide concentrations in the area.

Lotic Communities

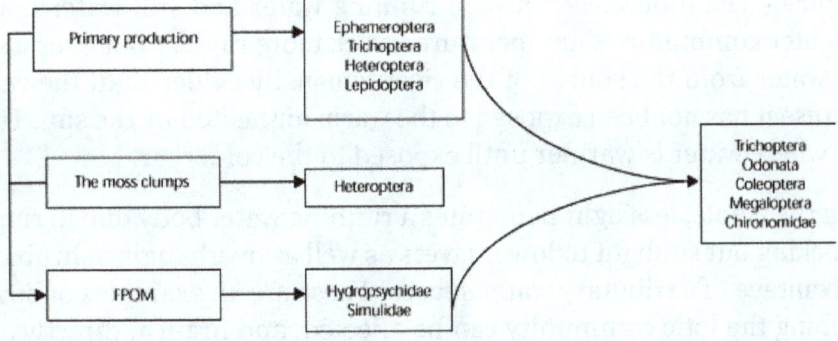

Running water freshwater communities are also known as lotic communities (lotic meaning running water). Lotic communities are formed by water being introduced to the freshwater body from a variety of sources;

- Rainfall - A percentage of water in the running water community will be present as a result of rainfall directly entering it.

- Ground Surface Water - Deriving from previous rainfall, water will enter the running water community.

- Underground Water - Water absorbed into the soil can also enter.

- Water Table - Deep underground there is a 'water table' which can also provide water for the running water community.

One of the main differences between lotic and lentic communities is the fact that the water is moving at a particular velocity in lotic communities. This can have great bearing on what organisms occupy the ecosystem and what particular ecological niche they can exist in. Running water can bring many factors into play affecting the lives of the organisms in this particular environment:

- Movement of minerals and stones caused by the velocity and volume of the water means the water bed is constantly changing. The faster and higher volume of water present will result in a direct increase in amount and size of particles shifted downstream.

- Standing waves are used by salmon at the bottom of waterfalls to spurn them upstream. At the same time, they cause small air pockets caused by oxygen replacing the splashing water, which results in a small micro-habitat becoming available suited to particular organisms.

- Erosion is caused by the running water breaking down the river bank and beds, causing the geography of the river to change over a long period of time. This means that hydroseres previously occupying the river bank may find themselves distanced from the running water for example, and over time this would mean the overall ecosystem would change over time.

The following is some of the physical and chemical factors that provide the framework of a running water community in which organisms in their favored ecological niches occupy.

- Temperature - The difference between running water and still water temperature is that running water communities' temperature varies more rapidly but over a smaller range. In summer, water from the source of the river is usually colder than the water found at the delta because it has not been exposed to the warm air heated by the sun. The reverse occurs in winter where water is warmer until exposed to the colder air.

- Light - On the whole, less light penetrates a running water body due to ripples in the water, debris blocking out sunlight to lower layers as well as overhanging shrubs that perhaps are taking advantage of a tributary water source. These are all examples of how the intensity of light reaching the lotic community can be affected, and in turn, directly affects the rate of photosynthesis done by plants in the community.

- Chemical Composition - Many factors can alter the chemical composition of the freshwater environment, including precipitation, the percolation of water via vegetation and sea spray to name a few. All in all, various elements and compounds are required by organisms in their daily activities and fluctuations or even an absence of such elements and compounds results in a direct effect on the lives of such organisms.

- Organic Matter - Organic matter previous external to the running water environment can also play a part in altering the ecosystem. This mostly occurs due to overhanging vegetation, although organic matter can be drawn into the ecosystem by the various sources mentioned on the previous page.

Lotic Communities & Algae

In general the diversity of plant species in a lotic community is small compared to that of a still water (lentic) community although small parts of the lotic community host similar conditions to that of a lentic community. Most plants have went through evolutionary adaptations to cope with the force and different conditions that running water brings. Such adaptations have allowed a number of species to successfully take advantage of the lotic community as their ecological niche.

As these conditions are more harsh for a typical species of plant, more notably larger plants, smaller species have found the conditions of the lotic community more favorable. This is due to the fact that they are more flexible in regards to the physical conditions of the water. Algae can grow in all sorts of different places and surfaces, and therefore are a successful constituent of the running water ecosystem. Most of these algae have developed evolutionary adaptations over times that prevent the water current sweeping them away.

There are many species of algae, all of which are capable of growing and reproducing at a quick rate. This consequence results in competition for niches in the freshwater environment, and in light of this, colonies of algae can heavily occupy one area at one moment in time and weeks later they can be succeeded by another species that can succeed in the conditions more favorably.

Algae are also the primary producers of this community, meaning they harness new energy into the ecosystem from the sun which provides the primary consumers with a valuable food source. With this in hand, it is apparent why algae populations and where they can be found in the lotic community is variable on a short-term basis.

Lotic Communities & Animals

The running water environment offers numerous microhabitats that simulate favorable conditions for many types of animals to successfully succeed the freshwater lotic community. As with plants, animals in this ecosystem have also undergone ongoing evolutionary adaptations to better suit this running water environment.

Some of these animals are sessile, meaning they are immobile and fixed to the one place. These animals are usually small, and include the protozoans and some freshwater sponges. These animals either remain attached to the mass of a plant or the water bank surface or rock. They usually obtain their food via tentacles which branch out into the flowing water and form a catchment area that can trap microscopic organisms (such as plankton) that is floating downstream.

As much as these sessile animals have developed adaptations to prevent being washed downstream, they are not thought to be one of the important pillars of the freshwater community. Over time when biotic and abiotic factors affect the landscape of the ecosystem over time, the location of these animals may not be as favorable as it once was, and they are unable to correct this due to their immobile nature. With this in light some animals have developed adaptations that allow them to travel through the water without being inhibited in same spot.

Animals have developed some of the following adaptations over time that helps them cope with the conditions in hand:

- Suckers - These suckers attach themselves to a surface that leeches them into position and can also assist movement in any given direction.

- Hooks / Claws - These sharp objects can dig into any given object and allow the animal to cling to a position or claw their way around the surface.

- Body flattening - This adaptation can allow the animal in the water bear less of the brunt of the force of water moving downstream, therefore reducing it as an inhibitor of their movement. This also allows these animals to enter confined areas (such as under stones) that may present a useful environment for them to live in.

- Streamlining - Just like man-made transport, animals who have underwent streamlining adaptations on their external appearance means that less resistance is presented by the running water when the animal attempts to move.

- Flight - Some animals have adaptations allowing them to fly, removing themselves from the force of the current at ground level and enabling them to move upstream more easily if needs be.

Ecology of Plankton

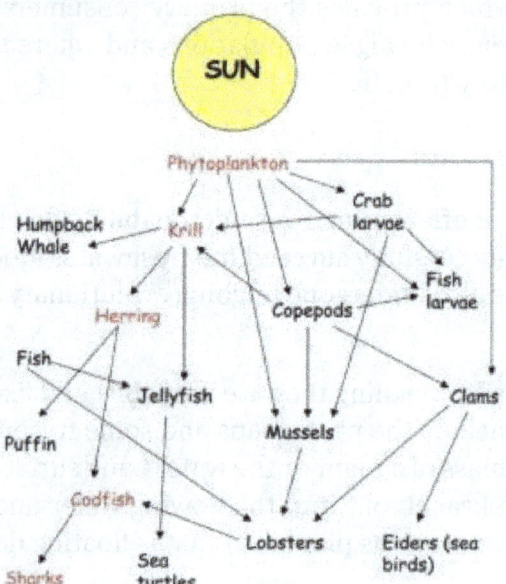

Plankton are those organisms which live suspended in the water of seas, lakes, ponds, and rivers, and which are not able to swim against the currents of water. This latter feature distinguishes plankton from nekton, the community of actively swimming organisms like fish, larger cephalopods, and aquatic mammals. Plankton range in size from ca. 0.2 gm to several meters (large jellyfish), but only the small ones have been the objects of intensive research, the Antarctic krill being the only well-studied plankton organism of > 5 mm.

Plankton form complex biotic communities which are functionally as diverse and show the same richness of interaction as terrestrial communities. Plankton are defined by their ecological niche rather than their phylogenetic or taxonomic classification. They provide a crucial source of food to larger, more familiar aquatic organisms such as fish.

The name plankton is derived from the Greek adjective planktos, meaning "errant", and by extension "wanderer" or "drifter". By definition, organisms classified as plankton are unable to resist ocean currents. While some forms are capable of independent movement and can swim hundreds of meters vertically in a single day (a behavior called diel vertical migration), their horizontal position is primarily determined by the surrounding currents. This is in contrast to nekton organisms that can swim against the ambient flow and control their position (e.g. squid, fish, and marine mammals).

Within the plankton, holoplankton spend their entire life cycle as plankton (e.g. most algae, copepods, salps, and some jellyfish). By contrast, meroplankton are only planktic for part of their lives (usually the larval stage), and then graduate to either the nekton or a benthic (sea floor) existence. Examples of meroplankton include the larvae of sea urchins, starfish, crustaceans, marine worms, and most fish.

Plankton abundance and distribution are strongly dependent on factors such as ambient nutrients concentrations, the physical state of the water column, and the abundance of other plankton. The study of plankton is termed planktology and individual plankton are referred to as plankters.

Functional groupings:

- Phytoplankton (from Greek phyton, or plant),

    Autotrophic, prokaryotic or eukaryotic algae that live near the water surface where there is sufficient light to support photosynthesis. Among the more important groups are the diatoms, cyanobacteria, dinoflagellates and coccolithophores.

- Zooplankton (from Greek zoon, or animal),

    Small protozoans or metazoans (e.g. crustaceans and other animals) that feed on other plankton and telonemia. Some of the eggs and larvae of larger animals, such as fish, crustaceans, and annelids.

- Bacterioplankton,

    Bacteria and archaea, which play an important role in remineralising organic material down the water column (note that the prokaryotic phytoplankton are also bacterioplankton).

This scheme divides the plankton community into broad producer, consumer and recycler groups. However, determining the trophic level of some plankton is not straightforward. For example, although most dinoflagellates are either photosynthetic producers or heterotrophic consumers, many species are mixotrophic depending upon their circumstances.

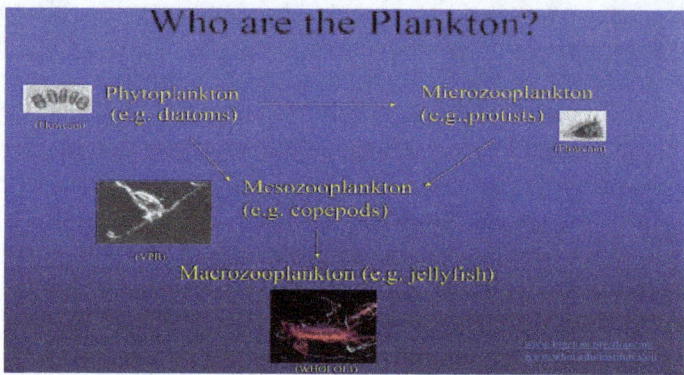

Distribution

Plankton inhabit oceans, seas and lakes. Local abundance varies horizontally, vertically and seasonally. The primary cause of this variability is the availability of light. All plankton ecosystems are driven by the input of solar energy (but seechemosynthesis), confining primary production to surface waters, and to geographical regions and seasons having abundant light.

A secondary variable is nutrient availability. Although large areas of the tropical and sub- tropical oceans have abundant light, they experience relatively low primary production because they offer limited nutrients such as nitrate, phosphate and silicate. This results from large-scale ocean circulation and water column stratification. In such regions, primary production usually occurs at greater depth, although at a reduced level (because of reduced light).

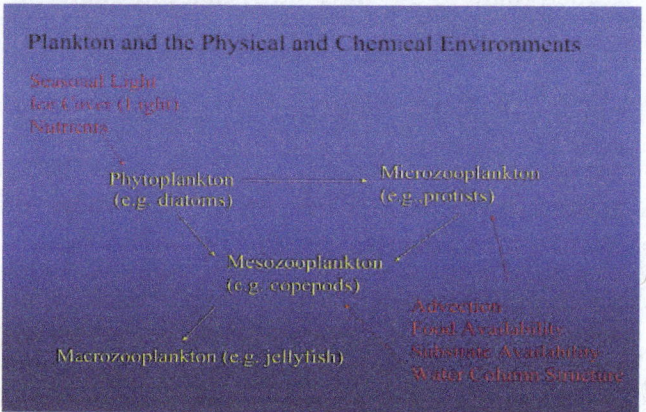

Despite significant macronutrient concentrations, some ocean regions are unproductive (so- called HNLC regions). The micronutrient iron is deficient in these regions, and adding it can lead to the formation of blooms of many kinds of phytoplankton. Iron primarily reaches the ocean through the deposition of dust on the sea surface. Paradoxically, oceanic areas adjacent to unproductive, arid land thus typically have abundant phytoplankton (e.g., the western Atlantic Ocean, where-trade winds bring dust from the Sahara Desert in north Africa). While plankton are most abundant in surface waters, they live throughout the water column. At depths where no primary production occurs, zooplankton and bacterioplankton instead consume organic material sinking from more productive surface waters above. This flux of sinking material, so-called marine snow, can be especially high following the termination of spring blooms.

Because of the central role of plankton in the functioning open-water foodwebs and ecosystems, plankton ecology has always been a core discipline of limnology and biological oceanography. Beyond their importance for aquatic systems, plankton are most suitable model organisms for classic topics of general ecology, such as competition, predator-prey relationships, food-web structure, succession, transfer of matter, and energy. Small size, rapid population growth (doubling times < 1 day for bacteria and small phytoplankton to several days or weeks for zooplankton), high abundances (millions per ml for bacteria, millions per 1 for phytoplankton), and a relatively homogeneous distribution in their environment facilitate field and experimental studies. Processes which take years to centuries in terrestrial systems, like competitive exclusion and succession, take only weeks in plankton.

Phytoplankton

The plantlike community of plankton is called phytoplankton, and the animal-like community is known as zooplankton. This convenient division is not without fault, for strictly speaking, many planktonic organisms are neither clearly plant nor animal but rather are better described as protists. When size is used as a criterion, plankton can be subdivided into macroplankton, microplankton, and nannoplankton, though no sharp lines can be drawn between these categories. Macroplankton can be collected with a coarse net, and morphological details of individual organisms are easily discernible. These forms, one millimetre or more in length, ordinarily do not include phytoplankton. Microplankton (also called net plankton) is composed of organisms between 0.05 and 1 mm in size and is a mixture of phytoplankton and zooplankton. The lower limit of its size range is fixed by the aperture of the finest cloth used for plankton nets. Nannoplankton (dwarf plankton) passes through all nets and consists of forms of a size less than 0.05 mm. Phytoplanktonic organisms dominate the nannoplankton.

Diatoms are one of the most common types of phytoplankton.

Phytoplankton are the autotrophic (self-feeding) components of the plankton community and a key part of oceans, seas and freshwater basin ecosystems. The name comes from the Greek words (*phyton*), meaning "plant", and (*planktos*), meaning "wanderer" or "drifter". Most phytoplankton are too small to be individually seen with the unaided eye. However, when present in high enough numbers, some varieties may be noticeable as colored patches on the water surface due to the presence of chlorophyll within their cells and accessory pigments (such as phycobiliproteins or xanthophylls) in some species.

Ecology

Phytoplankton are photosynthesizing microscopic organisms that inhabit the upper sunlit layer of almost all oceans and bodies of fresh water on Earth. They are agents for "primary production," the creation of organic compounds from carbon dioxide dissolved in the water, a process that sustains the aquatic food web.

Phytoplankton obtain energy through the process of photosynthesis and must therefore live in the well-lit surface layer (termed the euphotic zone) of an ocean, sea, lake, or other body of water. Phytoplankton account for about half of all photosynthetic activity on Earth. Their cumulative energy fixation in carbon compounds (primary production) is the basis for the vast majority of oceanic and also many freshwater food webs (chemosynthesis is a notable exception).

Phytoplankton come in many shapes and sizes.

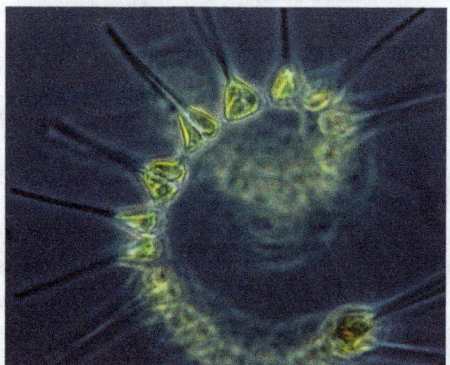

Phytoplankton are the foundation of the oceanic food chain.

The effects of anthropogenic warming on the global population of phytoplankton is an area of active research. Changes in the vertical stratification of the water column, the rate of temperature-dependent biological reactions, and the atmospheric supply of nutrients are expected to have important effects on future phytoplankton productivity. Additionally, changes in the mortality of phytoplankton due to rates of zooplankton grazing may be significant. As a side note, one of the more remarkable food chains in the ocean – remarkable because of the small number of links – is that of phytoplankton-feeding krill (a crustacean similar to a tiny shrimp) feeding baleen whales.

When two currents collide (here the Oyashio and Kuroshio currents) they create eddies.
Phytoplankton concentrates along the boundaries of the eddies, tracing the motion of the water.

Algal bloom off south England.

While almost all phytoplankton species are obligate photoautotrophs, there are some that are mixotrophic and other, non-pigmented species that are actually heterotrophic (the latter are often viewed as zooplankton). Of these, the best known are dinoflagellate genera such as *Noctiluca* and *Dinophysis*, that obtain organic carbon by ingesting other organisms or detrital material.

The term phytoplankton encompasses all photoautotrophic microorganisms in aquatic food webs. Phytoplankton serve as the base of the aquatic food web, providing an essential ecological function for all aquatic life. However, unlike terrestrial communities, where most autotrophs are plants, phytoplankton are a diverse group, incorporating protistan eukaryotes and both eubacterial and archaebacterial prokaryotes. There are about 5,000 known species of marine phytoplankton. How such diversity evolved despite scarce resources (restricting niche differentiation) is unclear.

In terms of numbers, the most important groups of phytoplankton include the diatoms, cyanobacteria and dinoflagellates, although many other groups of algae are represented. One group, the coccolithophorids, is responsible (in part) for the release of significant amounts of dimethyl sulfide (DMS) into the atmosphere. DMS is oxidized to form sulfate which, in areas where ambient aerosol particle concentrations are low, can contribute to the population of cloud condensation nuclei, mostly leading to increased cloud cover and cloud albedo according to the so-called CLAW Hypothesis. Different types of phytoplankton fill different trophic levels within varying ecosystems.

In oligotrophic oceanic regions such as the Sargasso Sea or the South Pacific Gyre, phytoplankton is dominated by the small sized cells, called picoplankton and nanoplankton (also referred to as picoflagellates and nanoflagellates), mostly composed of cyanobacteria (*Prochlorococcus*, *Synechococcus*) and picoeucaryotes such as *Micromonas*. Within more productive ecosystems, dominated by upwelling or high terrestrial inputs, larger dinoflagellates are the more dominant phytoplankton and reflect a larger portion of the biomass.

Minerals

Phytoplankton are crucially dependent on minerals. These are primarily macronutrients such as nitrate, phosphate or silicic acid, whose availability is governed by the balance between the so-called biological pump and upwelling of deep, nutrient-rich waters. However, across large regions of the World Ocean such as the Southern Ocean, phytoplankton are also limited by the lack of the micronutrient iron. This has led to some scientists advocating iron fertilization as a means to counteract the accumulation of human-produced carbon dioxide (CO_2) in the atmosphere. Large-scale experiments have added iron (usually as salts such as iron sulphate) to the oceans to promote phytoplankton growth and draw atmospheric CO_2 into the ocean. However, controversy about manipulating the ecosystem and the efficiency of iron fertilization has slowed such experiments.

Vitamin B

Phytoplankton depend on Vitamin B for survival. Areas in the ocean have been identified as having a major lack of Vitamin B, and correspondingly, phytoplankton.

Oxygen production

Phytoplankton absorb energy from the Sun and nutrients from the water to produce their own food. In the process of photosynthesis, phytoplankton release molecular oxygen (O_2) into the water. It is estimated that between 50% and 85% of the world's oxygen is produced via phytoplankton photosynthesis. The rest is produced via photosynthesis on land by plants. Furthermore, phytoplankton photosynthesis has controlled the atmospheric CO_2/O_2 balance since the early Precambrian Eon.

Growth Strategy

In the early twentieth century, Alfred C. Redfield found the similarity of the phytoplankton's elemental composition to the major dissolved nutrients in the deep ocean. Redfield proposed that the ratio of nitrogen to phosphorus (16:1) in the ocean was controlled by the phytoplankton's requirements which subsequently release nitrogen and phosphorus as they are remineralized. This so-called "Redfield ratio" in describing stoichiometry of phytoplankton and seawater has become a fundamental principle to understand the marine ecology, biogeochemistry and phytoplankton evolution. However, Redfield ratio is not a universal value and it may diverge due to the changes in exogenous nutrient delivery and microbial metabolisms in the ocean, such as nitrogen fixation, denitrification and anammox.

The dynamic stoichiometry shown in unicellular algae reflects their capability to stockpile nutrients in internal pool, shift between enzymes with various nutrient requirements and alter osmolyte composition. Different cellular components have their own unique stoichiometry characteristics, for instance, resource (light or nutrients) acquisition machinery such as proteins and chlorophyll contain high concentration of nitrogen but low in phosphorus. Meanwhile, growth machinery such as ribosomal RNA contains high nitrogen and phosphorus concentration.

Based on allocation of resources, phytoplankton is classified into three different growth strategies, namely survivalist, bloomer and generalist. Survivalist phytoplankton has high ratio of N:P (>30) and contains numerous resource-acquisition machinery to sustain growth under scarce resources. Bloomer phytoplankton has low N:P ratio (<10), contains high proportion of growth machinery and adapted to exponential growth. Generalist phytoplankton has similar N:P to Redfield ratio and contain relatively equal resource-acquisition and growth machinery.

Environmental Controversy

A 2010 study published in *Nature* reported that marine phytoplankton had declined substantially in the world's oceans over the past century. Phytoplankton concentrations in surface waters were estimated to have decreased by about 40% since 1950, at a rate of around 1% per year, possibly in response to ocean warming. The study generated debate among scientists and led to several communications and criticisms, also published in *Nature*. In a 2014 follow-up study, the authors used a larger database of measurements and revised their analysis methods to account for several of the published criticisms, but ultimately reached similar conclusions to the original *Nature* study. These studies and the need to understand the phytoplankon in the ocean led to the creation of the Secchi Disk Citizen Science study in 2013. The Secchi Disk study is a global study of phytoplankton conducted by seafarers (sailors, anglers, divers) involving a Secchi Disk and a smartphone app.

Estimates of oceanic phytoplankton change are highly variable. One global ocean primary productivity study found a net increase in phytoplankton, as judged from measured chlorophyll, when comparing observations in 1998–2002 to those conducted during a prior mission in 1979–1986. However, using the same database of measurements, other studies concluded that both chlorophyll and primary production had declined over this same time interval. The airborne fraction of CO_2 from human emissions, the percentage neither sequestered by photosynthetic life on land and sea nor absorbed in the oceans abiotically, has been almost constant over the past century, and that suggests a moderate upper limit on how much a component of the carbon cycle as large as phytoplankton have declined. In the northeast Atlantic, where a relatively long chlorophyll data series is available, and the site of the Continuous Plankton Recorder (CPR) survey, a net increase was found from 1948 to 2002. During 1998–2005, global ocean net primary productivity rose in 1998, followed by a decline during the rest of that period, yielding a small net increase. Using six climate model simulations, a large multi-university study of ocean ecosystems predicted that "a global increase in primary production of 0.7% at the low end to 8.1% at the high end," by 2050 although with "very large regional differences" including "a contraction of the highly productive marginal sea ice biome by 42% in the Northern Hemisphere and 17% in the Southern Hemisphere." A more recent multi-model study estimated that primary production would decline by 2-20% by 2100 A.D. Despite substantial variation in both the magnitude and spatial pattern of change, the majority of published studies predict that phytoplankton biomass and/or primary production will decline over the next century.

Researchers at the Woods Hole Oceanographic Institution have found phytoplankton to be a major source of methanol (CH_3OH) in the ocean in quantities that could rival or exceed that which is produced on land.

Aquaculture

Phytoplankton are a key food item in both aquaculture and mariculture. Both utilize phytoplankton as food for the animals being farmed. In mariculture, the phytoplankton is naturally occurring and is introduced into enclosures with the normal circulation of seawater. In aquaculture, phytoplankton must be obtained and introduced directly. The plankton can either be collected from a body of water or cultured, though the former method is seldom used. Phytoplankton is used as a foodstock for the production of rotifers, which are in turn used to feed other organisms. Phytoplankton is also used to feed many varieties of aquacultured molluscs, including pearl oysters and giant clams.

The production of phytoplankton under artificial conditions is itself a form of aquaculture. Phytoplankton is cultured for a variety of purposes, including foodstock for other aquacultured organisms, a nutritional supplement for captive invertebrates in aquaria. Culture sizes range from small-scale laboratory cultures of less than 1L to several tens of thousands of liters for commercial aquaculture. Regardless of the size of the culture, certain conditions must be provided for efficient growth of plankton. The majority of cultured plankton is marine, and seawater of a specific gravity of 1.010 to 1.026 may be used as a culture medium. This water must be sterilized, usually by either high temperatures in an autoclave or by exposure to ultraviolet radiation, to prevent biological contamination of the culture. Various fertilizers are added to the culture medium to facilitate the growth of plankton. A culture must be aerated or agitated in some way to keep plankton suspended, as well as to provide dissolved carbon dioxide for photosynthesis. In addition to constant aeration, most cultures are manually mixed or stirred on a regular basis. Light must be provided for the growth of phytoplankton. The colour temperature of illumination should be approximately 6,500 K, but values from 4,000 K to upwards of 20,000 K have been used successfully. The duration of light exposure should be approximately 16 hours daily; this is the most efficient artificial day length.

Zooplankton

A variety of zooplankton organisms

Zooplankton are heterotrophic (sometimes detritivorous) plankton. Plankton are organisms drifting in oceans, seas, and bodies of fresh water. The word "zooplankton" is derived from the Greek *zoon*, meaning animal, and *planktos*, meaning wanderer or drifter. Individual zooplankton are usually microscopic, but some (such as jellyfish) are larger and visible with the naked eye.

Ecology

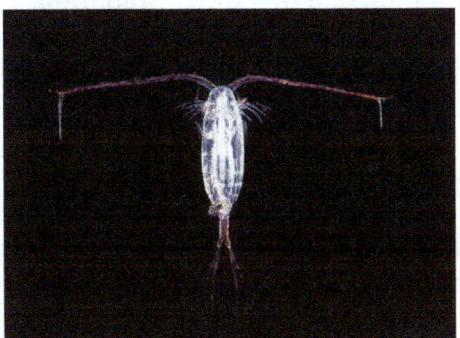

A copepod (*Calanoida* sp.)

A jellyfish (*Aequorea victoria*)

Zooplankton is a categorization spanning a range of organism sizes including small protozoans and large metazoans. It includes holoplanktonic organisms whose complete life cycle lies within the plankton, as well as meroplanktonic organisms that spend part of their lives in the plankton before graduating to either the nekton or a sessile, benthic existence. Although zooplankton are primarily transported by ambient water currents, many have locomotion, used to avoid predators (as in diel vertical migration) or to increase prey encounter rate.

Ecologically important protozoan zooplankton groups include the foraminiferans, radiolarians and dinoflagellates (the last of these are often mixotrophic). Important metazoan zooplankton include cnidarians such as jellyfish and the Portuguese Man o' War; crustaceans such as copepods, ostracods, isopods, amphipods, mysids and krill; chaetognaths (arrow worms); molluscs such as pteropods; and chordates such as salps and juvenile fish. This wide phylogenetic range includes a similarly wide range in feeding behavior: filter feeding, predation and symbiosis with autotrophic phytoplankton as seen in corals. Zooplankton feed on bacterioplankton, phytoplankton, other zooplankton (sometimes cannibalistically), detritus (or marine snow) and even nektonic organisms. As a result, zooplankton are primarily found in surface waters where food resources (phytoplankton or other zooplankton) are abundant.

Just as any species can be limited within a geographical region, so is zooplankton. However, species of zooplankton are not dispersed uniformly or randomly within a region of the ocean. As with phytoplankton, 'patches' of zooplankton species exist throughout the ocean. Though few physical barriers exist above the mesopelagic, specific species of zooplankton are strictly restricted by salinity and temperature gradients; while other species can withstand wide temperature and salinity gradients. Zooplankton patchiness can also be influenced by biological factors, as well as other physical factors. Biological factors include breeding, predation, concentration of phytoplankton, and vertical migration. The physical factor that influences zooplankton distribution the most is mixing of the water column (upwelling and downwelling along the coast and in the open ocean) that affects nutrient availability and, in turn, phytoplankton production.

Through their consumption and processing of phytoplankton and other food sources, zooplankton play a role in aquatic food webs, as a resource for consumers on higher trophic levels (including fish), and as a conduit for packaging the organic material in the biological pump. Since they are typically small, zooplankton can respond rapidly to increases in phytoplankton abundance, for instance, during the spring bloom.

Zooplankton can also act as a disease reservoir. Crustacean zooplankton have been found to house the bacterium *Vibrio cholerae*, which causes cholera, by allowing the cholera vibrios to attach to their chitinous exoskeletons. This symbiotic relationship enhances the bacterium's ability to survive in an aquatic environment, as the exoskeleton provides the bacterium with carbon and nitrogen.

Bacteria and Fungi

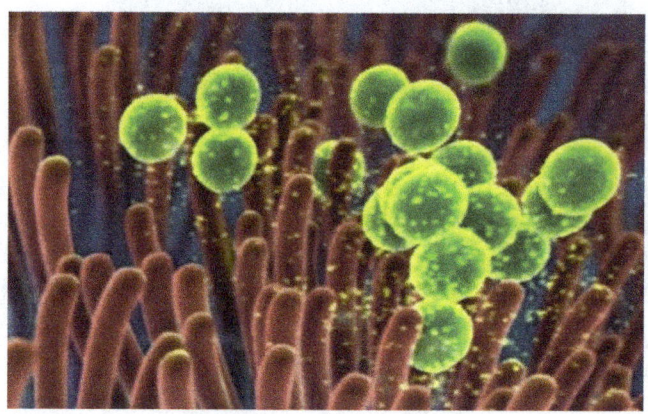

Bacteria and fungi found in water belong by definition to plankton, but, because of special techniques required for sampling and identification, they usually are considered separately. These organisms are important in the transformation of dead organic materials to inorganic plant nutrients. Some of these marine and freshwater microorganisms (including blue-green algae) fix molecular nitrogen from water containing dissolved air, forming ammonia or related nutrients important for phytoplankton growth. Although little is known about the extent of nitrogen fixation, bacteria and fungi always are found in water samples. A peculiar situation exists in the Black Sea, where water below 130–180 m contains hydrogen sulfide and no oxygen. Under these conditions only bacteria are found.

Ecosystems consist of populations, which in turn consist of individuals that interact with one another and with the environment. Biological interactions in the ocean are not between populations or between trophic levels, as many box-model representations of pelagic food webs might lead us to think. Trophic levels and populations are abstractions, and interactions occur at the level of the individual. —Blind sampling of bulk properties may result in observed distributional patterns, for example, that cannot be understood and explained from such an approach on its own. The picture must be complemented by approaches that consider the individual in its immediate environment and that provide a mechanistic understanding of the functioning of individuals and of components of the larger systems.

This allows us to build models and to extrapolate observations beyond the system in which the observations were made. Traditionally, scientists who go on cruises and examine distribution patterns of both biota and environmental properties using sampling are considered biological oceanographers, and those who explore the functioning of individuals, for example by conducting laboratory experiments with organisms, are considered marine biologists. We need to combine the two approaches to understand the ecology of the oceans.

The motivation to try to understand the ecology of planktonic organisms is twofold. The first driving force has to do with a simple interest in natural history. It is fascinating to watch the behavior of live plankters under the microscope or—better—free- swimming plankters by video; they have different but often beautiful forms and colors, and even closely related species may behave very differently, which makes identifying live plankton much easier than identifying dead ones. The second reason for examining the adaptations and behavior of plankters is our interest in understanding overall properties of pelagic systems and how the pelagic system relates to the larger- scale issues of fisheries' yield, CO_2 balance, global climate, and others. Understanding the mechanistics of individual behaviors and interactions may allow us to predict rates and to scale rates to sizes, which, in turn, may help us understand the (size) structure and function of pelagic systems and to predict effects of environmental changes and human impacts.

The Encounter Problem

Life is all about encounters. In the ocean, for example, phytoplankton cells need to encounter molecules of nutrient salts and inorganic carbon; bacteria need to encounter organic molecules; viruses need to encounter their hosts; predators need to encounter their prey; and males need to encounter females (or vice versa). Other important processes in the ocean, such as the formation of marine snow aggregates, likewise depend on encounters, here encounters between the component particles.

All organisms, including plankters, have three main tasks in life, namely to eat, to reproduce, and to avoid being eaten, all related to encounters or avoiding encounters. The behavior, morphology, and ecology of planktonic organisms must to a large extent represent adaptations to undertake these missions, and the diversity of form, function, and behavior that we can observe among plankters must be the result of different ways of solving the problems in the environment in which they live. The pelagic environment seen from the point of view of a small plankter is very different from the environment experienced by humans, and our intuition is often insufficient to allow us to understand the behavioural adaptations of planktonic organisms. Thus, although ornithologists to a large extent may be able to understand the behavior of their study organisms by using common sense, copepodologists rarely can, to rephrase the title of a classical ecology paper (Hutchinson 1951). For example, at the scale of planktonic organisms, the medium is viscous, and inertial forces therefore are insignificant, which makes moving an entirely different undertaking than what we as humans are used to or have seen other terrestrial animals do; the density of water is orders of magnitude higher than the density of air, which makes floatation easier and currents more important; for the smallest pelagic organisms (bacteria), thermally driven Brownian motion makes steering impossible; and most plankton use senses different from, and less far-reaching than, vision to perceive the environment. In addition, the pelagic environment is three- dimensional, whereas humans mainly move in only two dimensions. This implies, among other things, which average distances between a planktonic organism and its target may be very large, maybe thousands of body lengths. Because of the often non-intuitive nature of the immediate environment of small pelagic organisms, we need to appeal to fluid dynamic considerations in order to achieve a mechanistic understanding of the small- scale interactions between plankters and their environment.

In pursuing the encounter problem we can write a very general equation that describes encounter rates

$$E = \beta C_1 C_2$$

where E is the number of encounters happening per unit time and volume between particle types 1 and 2, C_1 and C_2 are the concentrations of these particles, and β is the encounter rate kernel ($L_3T - 1$). Often we are interested in looking at the per capita rate, that is, the rate at which one particle of type 1 encounters a particle of type 2:

$$e = E/C_1 = \beta C_2$$

For example, if particle 1 is a suspension-feeding ciliate and C_2 the concentration of its phytoplankton prey, then β is the ciliate's clearance rate, and e its ingestion rate (assuming that all encountered particles are ingested). The clearance rate is the equivalent volume of water from which the ciliate removes all prey particles per unit time. In many suspension- feeding ciliates, the clearance rate can be interpreted directly as a filtration rate; that is, the rate at which water is passed through a filtering structure that retains suspended particles. As a different but similar example: if particle 1 is a fish larva looking for food, and particle 2 its microzooplankton prey, then β is the volume of water that the larvae can search for prey items per unit time; if all encountered prey are consumed, then e is the ingestion rate of the fish larva. We may also see the process from the point of view of the prey, in which case βC_1 is the mortality rate of the phytoplankton or microzooplankton prey population through ciliate grazing or fish larval feeding. As a final example: if C_1 is the concentration of bacteria, and C_2 the concentration of organic molecules on which the bacteria feed, then e is the assimilation rate; it is more difficult to give a physical interpretation of β in this case. However, it is, like a clearance rate, the imaginary volume of water from which the bacterium removes all molecules per unit time. In fact, any encounter problem can be cast in terms of the general equation (eq. 1), but obviously the interpretation or meaning of the terms may be very different. The processes or mechanisms responsible for encounters are contained in the encounter-rate kernel. Obviously, from the examples above, these mechanisms are diverse. Intuitively, encounter rates depend on two factors: the motility of the encountering —particles‖ and the ambient fluid motion that may enhance encounter rates. Motility encompasses here the diffusivity of molecules, the swimming of organisms, and the sinking of particles. In regard to planktonic organisms, ambient fluid motion essentially means turbulence because planktonic organisms (contrary to benthic ones) are embedded in the general flow. From this consideration, one can see that there may be different components entering into the encounter-rate kernel depending on the specific problem under consideration.

Plankton and Biological Productivity

Plankton is the productive base of both marine and freshwater ecosystems, providing food for larger animals and indirectly for humans, whose fisheries depend upon plankton. The productivity of an area is dependent upon the availability of nutrients and water-stability conditions. Currents that flow near continents are important to plankton production in an area. The California Current (a continuation of the Kuroshio Drift from Japan) causes an outland transport of water and combines with a compensating nutrient-rich current along the coast of California to make this area highly productive. The same situation exists along the west coast of southern Africa, which is influenced by the Benguela Current, and off the west coast of South America, influenced by the Peru Current.

In the sea an adequate supply of nutrients, including carbon dioxide, enables phytoplankton and benthic algae to transform the light energy of the Sun into energy-rich chemical components through photosynthesis. The bottom-dwelling algae are responsible for about 2 percent of the primary production in the ocean; the remaining 98 percent is attributable to phytoplankton. Most of the phytoplankton serves as food for zooplankton, but some of it is carried below the light zone. After death, this phytoplankton undergoes chemical mineralization, bacterial breakdown, or transformation into sediments. Phytoplankton production usually is greatest from 5 to 10 m below the surface of the water. High light intensity and the lack of nutrient in the regions above a depth of 5 m may be the causes for suboptimal photosynthesis. Although bacteria are found at all depths, they are most abundant either immediately below great phytoplankton populations or just above the bottom.

As a human resource, plankton has only begun to be developed and exploited. It may in time be the chief food supply of the world, in view of its high biological productivity and wide extent. It has been demonstrated on several occasions that large-scale cultures of algae are technically feasible. The unicellular green alga Chorella has been used particularly in this connection. Through ample culture conditions, production is directed toward protein content greater than 50 percent. Although this protein has a suitable balance of essential amino acids, its low degree of digestibility prevents practical use. Phytoplankton may become increasingly important in space travel as a source for food and for gas exchange. The carbon dioxide released during respiration of spacecraft personnel would be transformed into organic substances by the algae, while the oxygen liberated during this process would support human respiration.

References

- "Major Source of Methanol in the Ocean Identified". Woods Hole Oceanographic Institution. March 10, 2016. Retrieved 2016-03-30

- Giller, S.; B. Malmqvist (1998). The Biology of Streams and Rivers. Oxford University Press, Oxford. p. 296. ISBN 0198549776

- Browne, R. A. (1981). "Lakes as islands: biogeographic distribution, turnover rates, and species composition in the lakes of central New York". Journal of Biogeography. 8 1: 75–83. JSTOR 2844594. doi:10.2307/2844594

- Ghosal; Rogers; Wray, S.; M.; A. "The Effects of Turbulence on Phytoplankton". Aerospace Technology Enterprise. NTRS. Retrieved 16 June 2011

- Hillebrand, H.; A. I. Azovsky (2001). "Body size determines the strength of the latitudinal diversity gradient". Ecography. 24 (3): 251–256. doi:10.1034/j.1600-0587.2001.240302.x

- Moss, B. (1998). Ecology of Freshwaters: man and medium, past to future. Blackwell Science, London. p. 557. ISBN 0632035129

- Calbet, A. (2008). "The trophic roles of microzooplankton in marine systems". ICES Journal of Marine Science. 65 (3): 325–31. doi:10.1093/icesjms/fsn013

- Roach, John (June 7, 2004). "Source of Half Earth's Oxygen Gets Little Credit". National Geographic News. Retrieved 2016-04-04

- Quinn, P. K.; Bates, T. S. (2011). "The case against climate regulation via oceanic phytoplankton sulphur emissions". Nature. 480 (7375): 51–6. PMID 22129724. doi:10.1038/nature10580

- Sterner, Robert Warner; Elser, James J. (2002). Ecological Stoichiometry: The Biology of Elements from Molecules to the Biosphere. Princeton University Press. ISBN 978-0-691-07491-7

- Antoine, David (2005). "Bridging ocean color observations of the 1980s and 2000s in search of long-term trends". Journal of Geophysical Research. 110 (C6). Bibcode:2005JGRC..110.6009A. doi:10.1029/2004JC002620

- Tappan, Helen (April 1968). "Primary production, isotopes, extinctions and the atmosphere". Palaeogeography, Palaeoclimatology, Palaeoecology. 4 (3): 187–210. doi:10.1016/0031-0182(68)90047-3. Retrieved 2016-04-04

- Arrigo, Kevin R. (2005). "Marine microorganisms and global nutrient cycles". Nature. 437 (7057): 349–55. PMID 16163345. doi:10.1038/nature04159

- Lalli, C.M. & Parsons, T.R. (1993). Biological Oceanography An Introduction. 30 Corporate Drive, Burlington, MA 01803: Elsevier. p. 314. ISBN 0-7506-3384-0

- Mackas, David L. (2011). "Does blending of chlorophyll data bias temporal trend?". Nature. 472 (7342): E4–5; discussion E8–9. PMID 21490623. doi:10.1038/nature09951

Fish Diversity: An Overview

Aquatic biodiversity is seen as the amount of living to nonliving matter in an aquatic region. 'Hotspots' are regions that possess a large amount of fish species. Aquatic biodiversity can be found in both freshwater and seawater environment. This section is an overview of the subject matter incorporating all the major aspects of aquaculture.

Fish Diversity

Ichthyodiversity refers to variety of fish species; depending on context and scale, it could refer to alleles or genotypes within piscian population, to species of life forms within a fish community, and to species of life forms across aquaregimes (Burton et.al., 1992). Biodiversity is also essential for stabilization of ecosystems, protection of overall environmental quality, for understanding intrinsic worth of all species on the earth (Ehrlich & Wilson, 1991). Positive correlations between biomass production and species abundance have been recorded in various earlier studies (Nikolsky, 1978). The species diversity of an ecosystem is often related to the amount of living and nonliving organic matter present in it. However, apparently, species diversity depends less on the characteristics of a single ecosystem than on the interaction between ecosystems, e.g., transport of living animals across the different gradient zones in the waterbody. The effect of such transport is an important 'information' exchange enhancing the genetic diversity. The genetic imprinting of various populations of lentic fish species is essential since the freshwater ecosystems constitute crucial parts of their life-support systems by providing nursing grounds and feeding areas (Hammer et al., 1993). Further, species diversity is a property at the population level while the functional diversity concept is more strongly related to ecosystem stability and stress, physical and chemical factors for determining population dynamics in the lentic ecosystem. Also, the various organisms including the plankton play a significant role in the dynamics of the ecosystem (Kar & Barbhuiya, 2004).

Fish constitutes almost half of the total number of vertebrates in the world. They live in almost all conceivable aquatic habitats. They exhibit enormous diversity of size, shape and biology, and in the habitats they occupy. Of the 39,900 species of vertebrates in the world, Nelson (1984) estimated 21,723 extant species of fish under 4,044 genera, 445 families and 50 Orders in the world, compared to 21,450 extant tetrapods. Of these, 8,411 are freshwater species and 11,650 are marine. Other researchers, have arrived at different estimates, most of which range between 17,000 and 30,000 for the numbers of currently recognized fish species. The eventual number of living fish species may be close to 28,000 in the world. Day (1889) described 1418 species of fish under 342 genera from the British India. The fish fauna of the major tropical regions, Southern Asia, Africa, South and Central America are generally different with respect to genera; but, some families have members in two or all of the continents. In Southern Asia the predominant fish groups are the carps (Cyprinidae) and the cat fishes (Siluroidea) (Berra, 1981).

India is one of the megabiodiversity countries in the world and occupies the ninth position in terms of freshwater megabiodiversity (Mittermeier & Mittermeier, 1997). The Indian fish population represents 11.72% of species, 23.96% of genera, 57% of families and 80% of the global fishes. Out of the 2200 species so far listed, 73 (3.32%) belong to the cold freshwater regime, 544 (24.73%) to the warm fresh waters domain, 143 (6.50%) to the brackish waters and 1440 (65.45%) to the marine ecosystem. This bewildering ichthyodiversity of this region has been attracting many ichthyologists both from India and abroad. Concomitantly, the northeastern region of India was identified as a biodiversity hotspot by the World Conservation Monitoring Centre (WCMC, 1998). This rich diversity of this region could be assigned to certain reasons, notably, the geomorphology and the tectonics of this zone. The hills, and the undulating valleys of this area gives rise to large number of torrential hill streams, which lead to big rivers; and, finally, become part of the Ganga-Brahmaputra-Barak-Chindwin-Kolodyne-Gomati-Meghna system (Kar, 2005).

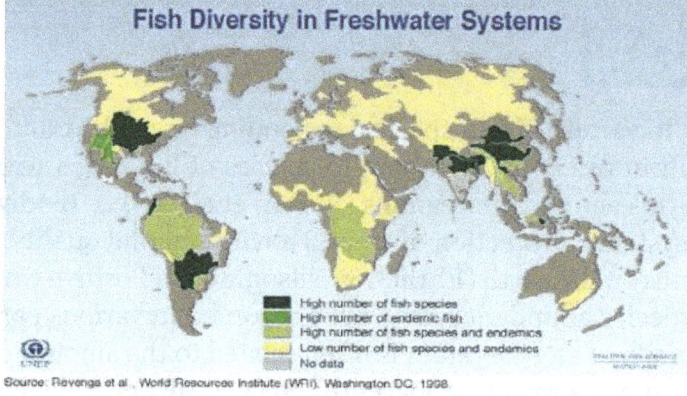

Source: Revenga et al., World Resources Institute (WRI), Washington DC, 1998.

The Indian fish fauna is divided into two classes, viz., Chondrichthyes and Osteichthyes. The Chondrichthyes are represented by 131 species under 67 genera, 28 families and 10 Orders in the Indian region. The annual average landings of the Indian Chondrichthyes is 33,442 tonnes, of which, 15,537 tonnes come from the east coast and 17,605 tonnes come from the west coast and the rest come from the Andaman and Nicobar, and Lakshadeep Islands.

The Indian Osteichthyes are represented by 2,415 species belonging to 902 genera, 226 families and 30 orders, of which, five families, notably the family Parapsilorhynchidae are endemic to India. These small hillstream fishes include a single genus, viz., Parapsilorhynchus which contains 3 species. They occur in the Western Ghats, Satpura mountains and the Bailadila range in Madhya Pradesh only. Further, the fishes of the family Psilorhynchidae with the only genus Psilorhynchus are also endemic to the Indian region. Other fishes endemic to India include the genus Olytra and the species Horaichthys setnai belonging to the families Olyridae and Horaichthyidae respectively. The latter occur from the Gulf of Kutch to Trivandrum coast. The endemic fish families form 2.21 per cent of the total bony fish families of the Indian region. 223 endemic fish species are found in India, representing 8.75 per cent of the total fish species known from the Indian region and 128 monotypic genera of fishes found in India, representing 13.20 per cent of the genera of fishes known from the Indian region.

There are about 450 families of freshwater fishes globally. Roughly 40 are represented in India (warm freshwater species). About 25 of these families contain commercially important species. Number of endemic species in warm water is about 544. Major warm water species are:

Bagarius bagarius, Catla catla, Channa marulius, C. punctatus, C. striatus, Cirrhinus mrigala, Clarias batrachus, Heteropneustes fossilis, Labeo bata, L. calbasu, L. rohita, Aorichthys seenghala, Notopterus chitala, N. notopterus, Pangasius pangasius, Rita rita, Wallago attu.

Cyprinids (family: Cyprinidae), Live fish (family: Anabantidae, Clariidae, Channidae, Heteropneustidae), Cat fish (family: Bagridae, Silurdae, Schilbeidae), Clupeids (family: Clupeidae), Mullets (family: Mugilidae), featherbacks (family: Notopteridae), Loaches (family: Cobitidae), Eels (family: Mastacembelidae), Glass fishes (family: Chandidae) and Gobies (family: Gobiidae) are the major groups of fresh water fishes found in India.

The Western Ghats, one of the well-known biodiversity hotspots of the world, harbours 289 species of freshwater fish of which 119 are endemic. The ecosystems in this region have been, over the past 150 years or so, experiencing tumultuous changes due to the ever-increasing human impacts. In this regard, a study was conducted in Sharavathi River, central Western Ghats to understand fish species composition with respect to landscape dynamics. The study, using a combination of remote-sensing data as well as field investigations shows that the streams having their catchments with high levels of ever greenness and endemic tree species of the Western Ghats were also richer in fish diversity and endemism, compared to those catchments with other kinds of vegetation. This illustrates that the composition and distribution of fish species have a strong association with the kind of terrestrial landscape elements and the importance of landscape approach to conservation and management of aquatic ecosystems.

Fish come in many shapes and sizes. This is a sea dragon, a close relative of the seahorse. They are camouflaged to look like floating seaweed.

The deep sea *Lasiognathus amphirhamphus* is a rare ambush predator known only from a single female specimen *(pictured)*. It is an angler fish that "angles" for its prey with a lure attached to a line from its head.

Fish are very diverse animals and can be categorised in many ways. This article is an overview of some of ways in which fish are categorised. Although most fish species have probably been discovered and described, about 250 new ones are still discovered every year. According to FishBase, 33,100 species of fish had been described by April 2015. That is more than the combined total of all other vertebrate species: mammals, amphibians, reptiles and birds.

Fish species diversity is roughly divided equally between marine (oceanic) and freshwater ecosystems. Coral reefs in the Indo-Pacific constitute the centre of diversity for marine fishes, whereas continental freshwater fishes are most diverse in large river basins of tropical rainforests, especially the Amazon, Congo, and Mekong basins. More than 5,600 fish species inhabit Neotropical freshwaters alone, such that Neotropical fishes represent about 10% of all vertebrate species on the Earth. Exceptionally rich sites in the Amazon basin, such as Cantão State Park, can contain more freshwater fish species than occur in all of Europe.

Coral Reef

Coral reefs are diverse underwater ecosystems held together by calcium carbonate structures secreted by corals. Coral reefs are built by colonies of tiny animals found in marine water that contain few nutrients. Most coral reefs are built from stony corals, which in turn consist of polyps that cluster in groups. The polyps belong to a group of animals known as Cnidaria, which also includes sea anemones and jellyfish. Unlike sea anemones, corals secrete hard carbonate exoskeletons which support and protect the coral polyps. Most reefs grow best in warm, shallow, clear, sunny and agitated water.

Often called "rainforests of the sea", shallow coral reefs form some of the most diverse ecosystems on Earth. They occupy less than 0.1% of the world's ocean surface, about half the area of France, yet they provide a home for at least 25% of all marine species, including fish, mollusks, worms, crustaceans, echinoderms, sponges, tunicates and other cnidarians. Paradoxically, coral reefs flourish even though they are surrounded by ocean waters that provide few nutrients. They are most commonly found at shallow depths in tropical waters, but deep water and cold water corals also exist on smaller scales in other areas.

Coral reefs deliver ecosystem services to tourism, fisheries and shoreline protection. The annual global economic value of coral reefs is estimated between US$29.8-375 billion. However, coral reefs are fragile ecosystems, partly because they are very sensitive to water temperature. They are under threat from climate change, oceanic acidification, blast fishing, cyanide fishing for aquarium fish, sunscreen use, overuse of reef resources, and harmful land-use practices, including urban and agricultural runoff and water pollution, which can harm reefs by encouraging excess algal growth.

Formation

Most of the coral reefs we can see today were formed after the last glacial period when melting ice caused the sea level to rise and flood the continental shelves. This means that most modern coral reefs are less than 10,000 years old. As communities established themselves on the shelves, the reefs grew upwards, pacing rising sea levels. Reefs that rose too slowly could become drowned reefs. They are covered by so much water that there was insufficient light. Coral reefs are found in the deep sea away from continental shelves, around oceanic islands and as atolls. The vast majority of these islands are volcanic in origin. The few exceptions have tectonic origins where plate movements have lifted the deep ocean floor on the surface.

In 1842 in his first monograph, *The Structure and Distribution of Coral Reefs*, Charles Darwin set out his theory of the formation of atoll reefs, an idea he conceived during the voyage of the *Beagle*. He theorized uplift and subsidence of the Earth's crust under the oceans formed the atolls. Darwin's theory sets out a sequence of three stages in atoll formation. It starts with a fringing reef forming around an extinct volcanic island as the island and ocean floor subsides. As the subsidence continues, the fringing reef becomes a barrier reef, and ultimately an atoll reef.

Darwin's theory starts with a volcanic island which becomes extinct

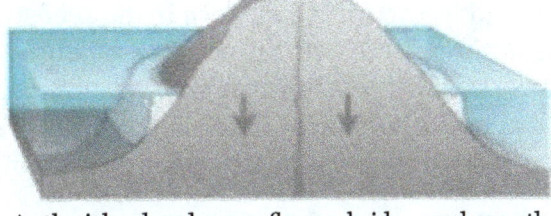

As the island and ocean floor subside, coral growth builds a fringing reef, often including a shallow lagoon between the land and the main reef.

As the subsidence continues, the fringing reef becomes a larger barrier reef further from the shore with a bigger and deeper lagoon inside.

Ultimately, the island sinks below the sea, and the barrier reef becomes an atoll enclosing an open lagoon.

Darwin predicted that underneath each lagoon would be a bed rock base, the remains of the original volcano. Subsequent drilling proved this correct. Darwin's theory followed from his understanding that coral polyps thrive in the clean seas of the tropics where the water is agitated, but can only live within a limited depth range, starting just below low tide. Where the level of the underlying earth allows, the corals grow around the coast to form what he called fringing reefs, and can eventually grow out from the shore to become a barrier reef.

Where the bottom is rising, fringing reefs can grow around the coast, but coral raised above sea level dies and becomes white limestone. If the land subsides slowly, the fringing reefs keep pace by growing upwards on a base of older, dead coral, forming a barrier reef enclosing a lagoon between the reef and the land. A barrier reef can encircle an island, and once the island sinks below sea level a roughly circular atoll of growing coral continues to keep up with the sea level, forming a central lagoon. Barrier reefs and atolls do not usually form complete circles, but are broken in places by storms. Like sea level rise, a rapidly subsiding bottom can overwhelm coral growth, killing the coral polyps and the reef, due to what is called *coral drowning*. Corals that rely on zooxanthellae can *drown* when the water becomes too deep for their symbionts to adequately photosynthesize, due to decreased light exposure.

The two main variables determining the geomorphology, or shape, of coral reefs are the nature of the underlying substrate on which they rest, and the history of the change in sea level relative to that substrate.

The approximately 20,000-year-old Great Barrier Reef offers an example of how coral reefs formed on continental shelves. Sea level was then 120 m (390 ft) lower than in the 21st century. As sea level rose, the water and the corals encroached on what had been hills of the Australian coastal plain. By 13,000 years ago, sea level had risen to 60 m (200 ft) lower than at present, and many hills of the coastal plains had become continental islands. As the sea level rise continued, water topped most of the continental islands. The corals could then overgrow the hills, forming the present cays and reefs. Sea level on the Great Barrier Reef has not changed significantly in the last 6,000 years, and the age of the modern living reef structure is estimated to be between 6,000 and 8,000 years. Although the Great Barrier Reef formed along a continental shelf, and not around a volcanic island, Darwin's principles apply. Development stopped at the barrier reef stage, since Australia is not about to submerge. It formed the world's largest barrier reef, 300–1,000 m (980–3,280 ft) from shore, stretching for 2,000 km (1,200 mi).

Healthy tropical coral reefs grow horizontally from 1 to 3 cm (0.39 to 1.18 in) per year, and grow vertically anywhere from 1 to 25 cm (0.39 to 9.84 in) per year; however, they grow only at depths shallower than 150 m (490 ft) because of their need for sunlight, and cannot grow above sea level.

Materials

As the name implies, the bulk of coral reefs is made up of coral skeletons from mostly intact coral colonies. As other chemical elements present in corals become incorporated into the calcium carbonate deposits, aragonite is formed. However, shell fragments and the remains of calcareous algae such as the green-segmented genus *Halimeda* can add to the reef's ability to withstand damage from storms and other threats. Such mixtures are visible in structures such as Eniwetok Atoll.

Types

The three principal reef types are:

- Fringing reef – directly attached to a shore, or borders it with an intervening shallow channel or lagoon

- Barrier reef – reef separated from a mainland or island shore by a deep channel or lagoon

- Atoll reef – more or less circular or continuous barrier reef extends all the way around a lagoon without a central island

A small atoll in the Maldives

Inhabited cay in the Maldives

Other reef types or variants are:

- Patch reef – common, isolated, comparatively small reef outcrop, usually within a lagoon or embayment, often circular and surrounded by sand or seagrass

- Apron reef – short reef resembling a fringing reef, but more sloped; extending out and downward from a point or peninsular shore

- Bank reef – linear or semicircular shaped-outline, larger than a patch reef

- Ribbon reef – long, narrow, possibly winding reef, usually associated with an atoll lagoon

- Table reef – isolated reef, approaching an atoll type, but without a lagoon

- Habili – reef specific to the Red Sea; does not reach the surface near enough to cause visible surf; may be a hazard to ships (from the Arabic for "unborn")

- Microatoll – community of species of corals; vertical growth limited by average tidal height; growth morphologies offer a low-resolution record of patterns of sea level change; fossilized remains can be dated using radioactive carbon dating and have been used to reconstruct Holocene sea levels

- Cays – small, low-elevation, sandy islands formed on the surface of coral reefs from eroded material that piles up, forming an area above sea level; can be stabilized by plants to become habitable; occur in tropical environments throughout the Pacific, Atlantic and Indian Oceans (including the Caribbean and on the Great Barrier Reef and Belize Barrier Reef), where they provide habitable and agricultural land

- Seamount or guyot – formed when a coral reef on a volcanic island subsides; tops of sea-mounts are rounded and guyots are flat; flat tops of guyots, or *tablemounts*, are due to erosion by waves, winds, and atmospheric processes.

Zones

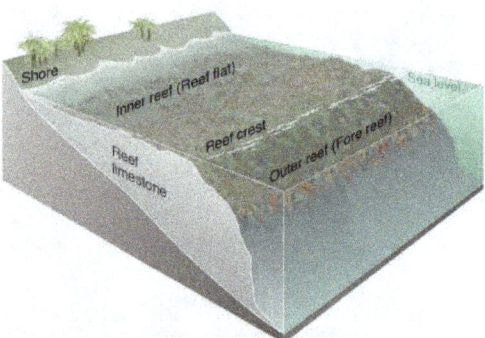

The three major zones of a coral reef: the fore reef, reef crest, and the back reef

Coral reef ecosystems contain distinct zones that represent different kinds of habitats. Usually, three major zones are recognized: the fore reef, reef crest, and the back reef (frequently referred to as the reef lagoon).

All three zones are physically and ecologically interconnected. Reef life and oceanic processes create opportunities for exchange of seawater, sediments, nutrients, and marine life among one another.

Thus, they are integrated components of the coral reef ecosystem, each playing a role in the support of the reefs' abundant and diverse fish assemblages.

Most coral reefs exist in shallow waters less than 50 m deep. Some inhabit tropical continental shelves where cool, nutrient rich upwelling does not occur, such as Great Barrier Reef. Others are found in the deep ocean surrounding islands or as atolls, such as in the Maldives. The reefs surrounding islands form when islands subside into the ocean, and atolls form when an island subsides below the surface of the sea.

Alternatively, Moyle and Cech distinguish six zones, though most reefs possess only some of the zones.

The reef surface is the shallowest part of the reef. It is subject to the surge and the rise and fall of tides. When waves pass over shallow areas, they shoal. This means the water is often agitated. These are the precise condition under which corals flourish. Shallowness means there is plenty of light for photosynthesis by the symbiotic zooxanthellae, and agitated water promotes the ability of coral to feed on plankton. However, other organisms must be able to withstand the robust conditions to flourish in this zone.

The off-reef floor is the shallow sea floor surrounding a reef. This zone occurs by reefs on continental shelves. Reefs around tropical islands and atolls drop abruptly to great depths, and do not have a floor. Usually sandy, the floor often supports seagrass meadows which are important foraging areas for reef fish.

The reef drop-off is, for its first 50 m, habitat for many reef fish who find shelter on the cliff face and plankton in the water nearby. The drop-off zone applies mainly to the reefs surrounding oceanic islands and atolls.

The reef face is the zone above the reef floor or the reef drop-off. This zone is often the most diverse area of the reef. Coral and calcareous algae growths provide complex habitats and areas which offer protection, such as cracks and crevices. Invertebrates and epiphytic algae provide much of the food for other organisms. A common feature on this forereef zone is spur and groove formations which serve to transport sediment downslope.

The reef flat is the sandy-bottomed flat, which can be behind the main reef, containing chunks of coral. This zone may border a lagoon and serve as a protective area, or it may lie between the reef and the shore, and in this case is a flat, rocky area. Fishes tend to prefer living in that flat, rocky area, compared to any other zone, when it is present.

The reef lagoon is an entirely enclosed region, which creates an area less affected by wave action that often contains small reef patches.

However, the "topography of coral reefs is constantly changing. Each reef is made up of irregular patches of algae, sessile invertebrates, and bare rock and sand. The size, shape and relative abundance of these patches changes from year to year in response to the various factors that favor one type of patch over another. Growing coral, for example, produces constant change in the fine structure of reefs. On a larger scale, tropical storms may knock out large sections of reef and cause boulders on sandy areas to move."

Locations

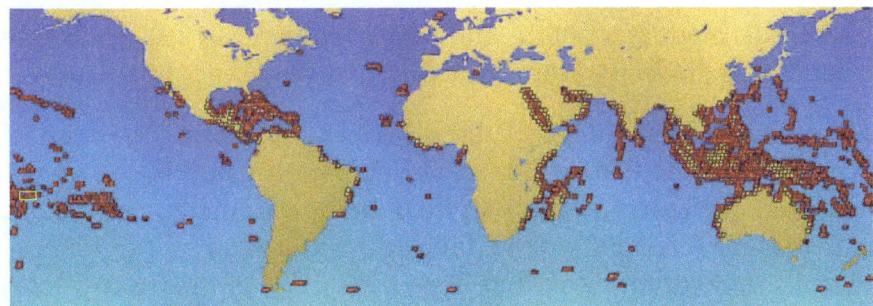

Locations of coral reefs

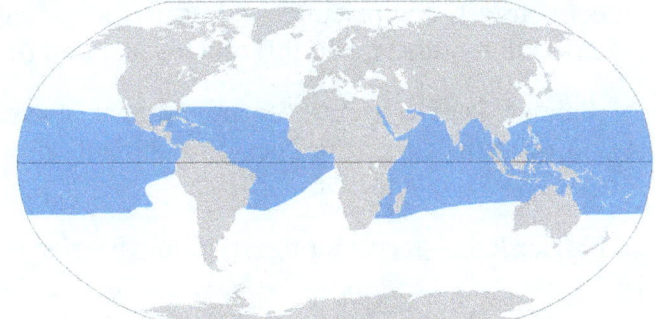

Boundary for 20 °C isotherms. Most corals live within this boundary. Note the cooler waters caused by upwelling on the southwest coast of Africa and off the coast of Peru.

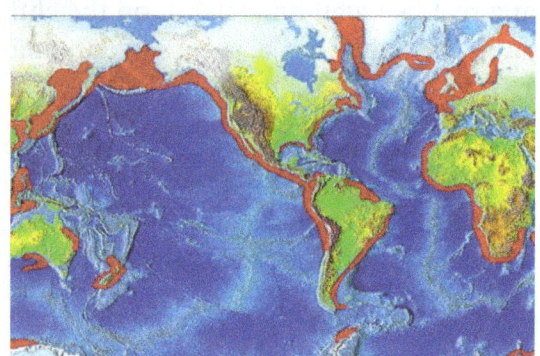

This map shows areas of upwelling in red. Coral reefs are not found in coastal areas where colder and nutrient-rich upwellings occur.

Marine biodiversity of Raja Ampat Islands, Indonesia. It is estimated that about 75% of world coral reef population are in the Raja Ampat Islands

Coral reefs are estimated to cover 284,300 km² (109,800 sq mi), just under 0.1% of the oceans' surface area. The Indo-Pacific region (including the Red Sea, Indian Ocean, Southeast Asia and the Pacific) account for 91.9% of this total. Southeast Asia accounts for 32.3% of that figure, while the Pacific including Australia accounts for 40.8%. Atlantic and Caribbean coral reefs account for 7.6%.

Although corals exist both in temperate and tropical waters, shallow-water reefs form only in a zone extending from approximately 30° N to 30° S of the equator. Tropical corals do not grow at depths of over 50 meters (160 ft). The optimum temperature for most coral reefs is 26–27 °C (79–81 °F), and few reefs exist in waters below 18 °C (64 °F). However, reefs in the Persian Gulf have adapted to temperatures of 13 °C (55 °F) in winter and 38 °C (100 °F) in summer. There are 37 species of scleractinian corals identified in such harsh environment around Larak Island.

Deep-water coral can exist at greater depths and colder temperatures at much higher latitudes, as far north as Norway. Although deep water corals can form reefs, very little is known about them.

Coral reefs are rare along the west coasts of the Americas and Africa, due primarily to upwelling and strong cold coastal currents that reduce water temperatures in these areas (respectively the Peru, Benguela and Canary streams). Corals are seldom found along the coastline of South Asia—from the eastern tip of India (Chennai) to the Bangladesh and Myanmar borders—as well as along the coasts of northeastern South America and Bangladesh, due to the freshwater release from the Amazon and Ganges Rivers respectively.

- The Great Barrier Reef—largest, comprising over 2,900 individual reefs and 900 islands stretching for over 2,600 kilometers (1,600 mi) off Queensland, Australia

- The Mesoamerican Barrier Reef System—second largest, stretching 1,000 kilometers (620 mi) from Isla Contoy at the tip of the Yucatán Peninsula down to the Bay Islands of Honduras

- The New Caledonia Barrier Reef—second longest double barrier reef, covering 1,500 kilometers (930 mi)

- The Andros, Bahamas Barrier Reef—third largest, following the east coast of Andros Island, Bahamas, between Andros and Nassau

- The Red Sea—includes 6000-year-old fringing reefs located around a 2,000 km (1,240 mi) coastline.

- The Florida Reef Tract—largest continental US reef, extends from Soldier Key, located in Biscayne Bay, to the Dry Tortugas in the Gulf of Mexico.

- Pulley Ridge—deepest photosynthetic coral reef, Florida.

- Numerous reefs scattered over the Maldives.

- The Philippines coral reef area, the second largest in Southeast Asia, is estimated at 26,000 square kilometers and holds an extraordinary diversity of species. Scientists have identified 915 reef fish species and more than 400 scleractinian coral species, 12 of which are endemic.

- The Raja Ampat Islands in Indonesia's West Papua province offer the highest known marine diversity.

- Bermuda is known for its northernmost coral reef system, located at 32.4° N and 64.8° W. The presence of coral reefs at this high latitude is due to the proximity of the Gulf Stream. Bermuda has a fairly consistent diversity of coral species, representing a subset of those found in the greater Caribbean.

- The world's northernmost individual coral reef so far discovered is located within a bay of Japan's Tsushima Island in the Korea Strait.

- The world's southernmost coral reef is at Lord Howe Island, in the Pacific Ocean off the east coast of Australia.

Biology

Alive corals are colonies of small animals embedded in calcium carbonate shells. It is a mistake to think of coral as plants or rocks. Coral heads consist of accumulations of individual animals called polyps, arranged in diverse shapes. Polyps are usually tiny, but they can range in size from a pinhead to 12 inches (30 cm) across.

Reef-building or hermatypic corals live only in the photic zone (above 50 m), the depth to which sufficient sunlight penetrates the water, allowing photosynthesis to occur. Coral polyps do not photosynthesize, but have a symbiotic relationship with microscopic algae of the genus *Symbiodinium*, commonly referred to as zooxanthellae. These organisms live within the tissues of polyps and provide organic nutrients that nourish the polyp. Because of this relationship, coral reefs grow much faster in clear water, which admits more sunlight. Without their symbionts, coral growth would be too slow to form significant reef structures. Corals get up to 90% of their nutrients from their symbionts.

Reefs grow as polyps and other organisms deposit calcium carbonate, the basis of coral, as a skeletal structure beneath and around themselves, pushing the coral head's top upwards and outwards. Waves, grazing fish (such as parrotfish), sea urchins, sponges, and other forces and organisms act as bioeroders, breaking down coral skeletons into fragments that settle into spaces in the reef structure or form sandy bottoms in associated reef lagoons. Many other organisms living in the reef community contribute skeletal calcium carbonate in the same manner. Coralline algae are important contributors to reef structure in those parts of the reef subjected to the greatest forces by waves (such as the reef front facing the open ocean). These algae strengthen the reef structure by depositing limestone in sheets over the reef surface.

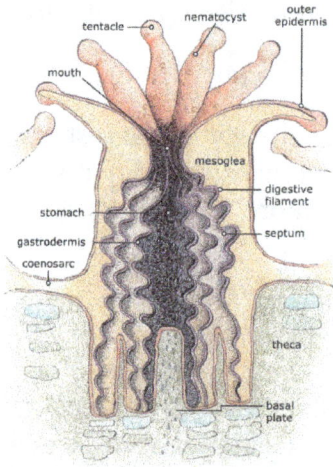

Anatomy of a coral polyp

Typical shapes for coral species are wrinkled brains, cabbages, table tops, antlers, wire strands and pillars. These shapes can depend on the life history of the coral, like light exposure and wave action, and events such as breakages.

Table coral

Close up of polyps are arrayed on a coral, waving their tentacles. There can be thousands of polyps on a single coral branch.

Corals reproduce both sexually and asexually. An individual polyp uses both reproductive modes within its lifetime. Corals reproduce sexually by either internal or external fertilization. The reproductive cells are found on the mesenteries, membranes that radiate inward from the layer of tissue that lines the stomach cavity. Some mature adult corals are hermaphroditic; others are exclusively male or female. A few species change sex as they grow.

Internally fertilized eggs develop in the polyp for a period ranging from days to weeks. Subsequent development produces a tiny larva, known as a planula. Externally fertilized eggs develop during synchronized spawning. Polyps release eggs and sperm into the water en masse, simultaneously. Eggs disperse over a large area. The timing of spawning depends on time of year, water temperature, and tidal and lunar cycles. Spawning is most successful when there is little variation between high and low tide. The less water movement, the better the chance for fertilization. Ideal timing occurs in the spring. Release of eggs or planula usually occurs at night, and is sometimes in phase with the lunar cycle (three to six days after a full moon). The period from release to settlement lasts only a few days, but some planulae can survive afloat for several weeks. They are vulnerable to predation and environmental conditions. The lucky few planulae which successfully attach to substrate next confront competition for food and space.

There are eight clades of Symbiodinium phylotypes. Most research has been completed on the Symbiodinium clades A–D. Each one of the eight contributes their own benefits as well as less compatible attributes to the survival of their coral hosts. Each photosynthetic organism has a specific level of sensitivity to photodamage of compounds needed for survival, such as proteins. Rates of regeneration and replication determine the organism's ability to survive. Phylotype A is found more in the shallow regions of marine waters. It is able to produce mycosporine-like amino acids that are UV resistant, using a derivative of glycerin to absorb the UV radiation and allowing them to become more receptive to warmer water temperatures. In the event of UV or thermal damage, if and when repair occurs, it will increase the likelihood of survival of the host and symbiont. This leads to the idea that, evolutionarily, clade A is more UV resistant and thermally resistant than the other clades.

Clades B and C are found more frequently in the deeper water regions, which may explain the higher susceptibility to increased temperatures. Terrestrial plants that receive less sunlight because they are found in the undergrowth can be analogized to clades B, C, and D. Since clades B through D are found at deeper depths, they require an elevated light absorption rate to be able to synthesize as much energy. With elevated absorption rates at UV wavelengths, the deeper occurring phylotypes are more prone to coral bleaching versus the more shallow clades. Clade D has been observed to be high temperature-tolerant, and as a result it has a higher rate of survival than clades B and C.

Brain coral

Staghorn coral

Mushroom coral   Maze coral

Darwin's Paradox

"Coral… seems to proliferate when ocean waters are warm, poor, clear and agitated, a fact which Darwin had already noted when he passed through Tahiti in 1842. This constitutes a fundamental paradox, shown quantitatively by the apparent impossibility of balancing input and output of the nutritive elements which control the coral polyp metabolism.

Recent oceanographic research has brought to light the reality of this paradox by confirming that the oligotrophy of the ocean euphotic zone persists right up to the swell-battered reef crest. When you approach the reef edges and atolls from the quasidesert of the open sea, the near absence of living matter suddenly becomes a plethora of life, without transition. So why is there something rather than nothing, and more precisely, where do the necessary nutrients for the functioning of this extraordinary coral reef machine come from?"

— Francis Rougerie

In *The Structure and Distribution of Coral Reefs*, published in 1842, Darwin described how coral reefs were found in some areas of the tropical seas but not others, with no obvious cause. The largest and strongest corals grew in parts of the reef exposed to the most violent surf and corals were weakened or absent where loose sediment accumulated.

Tropical waters contain few nutrients yet a coral reef can flourish like an "oasis in the desert". This has given rise to the ecosystem conundrum, sometimes called "Darwin's paradox": "How can such high production flourish in such nutrient poor conditions?"

Coral reefs cover less than 0.1% of the surface of the world's ocean, about half the land area of France, yet they support over one-quarter of all marine species. This diversity results in complex food webs, with large predator fish eating smaller forage fish that eat yet smaller zooplankton and so on. However, all food webs eventually depend on plants, which are the primary producers. Coral reefs' primary productivity is very high, typically producing 5–10 grams of carbon per square meter per day ($gC·m^{-2}·day^{-1}$) biomass.

One reason for the unusual clarity of tropical waters is they are deficient in nutrients and drifting plankton. Further, the sun shines year-round in the tropics, warming the surface layer, making it less dense than subsurface layers. The warmer water is separated from deeper, cooler water by a stable thermocline, where the temperature makes a rapid change. This keeps the warm surface waters floating above the cooler deeper waters. In most parts of the ocean, there is little exchange between these layers. Organisms that die in aquatic environments generally sink to the bottom, where they decompose, which releases nutrients in the form of nitrogen (N), phosphorus (P) and potassium (K). These nutrients are necessary for plant growth, but in the tropics, they do not directly return to the surface.

Plants form the base of the food chain, and need sunlight and nutrients to grow. In the ocean, these plants are mainly microscopic phytoplankton which drift in the water column. They need sunlight for photosynthesis, which powers carbon fixation, so they are found only relatively near the surface. But they also need nutrients. Phytoplankton rapidly use nutrients in the surface waters, and in the tropics, these nutrients are not usually replaced because of the thermocline.

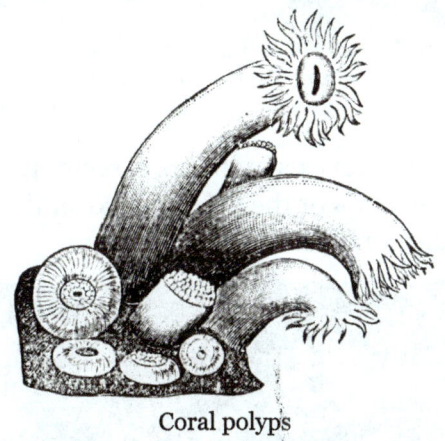

Coral polyps

Explanations

Around coral reefs, lagoons fill in with material eroded from the reef and the island. They become havens for marine life, providing protection from waves and storms.

Most importantly, reefs recycle nutrients, which happens much less in the open ocean. In coral reefs and lagoons, producers include phytoplankton, as well as seaweed and coralline algae, especially small types called turf algae, which pass nutrients to corals. The phytoplankton are eaten by fish and crustaceans, who also pass nutrients along the food web. Recycling ensures fewer nutrients are needed overall to support the community.

Coral reefs support many symbiotic relationships. In particular, zooxanthellae provide energy to coral in the form of glucose, glycerol, and amino acids. Zooxanthellae can provide up to 90% of a coral's energy requirements. In return, as an example of mutualism, the corals shelter the zooxanthellae, averaging one million for every cubic centimeter of coral, and provide a constant supply of the carbon dioxide they need for photosynthesis.

Corals also absorb nutrients, including inorganic nitrogen and phosphorus, directly from water. Many corals extend their tentacles at night to catch zooplankton that brush them when the water is agitated. Zooplankton provide the polyp with nitrogen, and the polyp shares some of the nitrogen with the zooxanthellae, which also require this element. The varying pigments in different species of zooxanthellae give them an overall brown or golden-brown appearance, and give brown corals their colors. Other pigments such as reds, blues, greens, etc. come from colored proteins made by the coral animals. Coral which loses a large fraction of its zooxanthellae becomes white (or sometimes pastel shades in corals that are richly pigmented with their own colorful proteins) and is said to be bleached, a condition which, unless corrected, can kill the coral.

The color of corals depends on the combination of brown shades provided by their zooxanthellae and pigmented proteins (reds, blues, greens, etc.) produced by the corals themselves.

Sponges are another key: they live in crevices in the coral reefs. They are efficient filter feeders, and in the Red Sea they consume about 60% of the phytoplankton that drifts by. The sponges eventually excrete nutrients in a form the corals can use.

Most coral polyps are nocturnal feeders. Here, in the dark, polyps have extended their tentacles to feed on zooplankton.

The roughness of coral surfaces is the key to coral survival in agitated waters. Normally, a boundary layer of still water surrounds a submerged object, which acts as a barrier. Waves breaking on the extremely rough edges of corals disrupt the boundary layer, allowing the corals access to passing nutrients. Turbulent water thereby promotes reef growth and branching. Without the nutritional gains brought by rough coral surfaces, even the most effective recycling would leave corals wanting in nutrients.

Studies have shown that deep nutrient-rich water entering coral reefs through isolated events may have significant effects on temperature and nutrient systems. This water movement disrupts the relatively stable thermocline that usually exists between warm shallow water to deeper colder water. Leichter et al. (2006) found that temperature regimes on coral reefs in the Bahamas and Florida were highly variable with temporal scales of minutes to seasons and spatial scales across depths.

Water can be moved through coral reefs in various ways, including current rings, surface waves, internal waves and tidal changes. Movement is generally created by tides and wind. As tides interact with varying bathymetry and wind mixes with surface water, internal waves are created. An internal wave is a gravity wave that moves along density stratification within the ocean. When a water parcel encounters a different density it will oscillate and create internal waves. While internal waves generally have a lower frequency than surface waves, they often form as a single wave that breaks into multiple waves as it hits a slope and moves upward. This vertical break up of internal waves causes significant diapycnal mixing and turbulence. Internal waves can act as nutrient pumps, bringing plankton and cool nutrient-rich water up to the surface.

The irregular structure characteristic of coral reef bathymetry may enhance mixing and produce pockets of cooler water and variable nutrient content. Arrival of cool, nutrient-rich water from depths due to internal waves and tidal bores has been linked to growth rates of suspension feeders and benthic algae as well as plankton and larval organisms. Leichter et al. proposed that Codium isthmocladum react to deep water nutrient sources due to their tissues having different concentrations of nutrients dependent upon depth. Wolanski and Hamner noted aggregations of eggs, larval organisms and plankton on reefs in response to deep water intrusions. Similarly, as internal waves and bores move vertically, surface-dwelling larval organisms are carried toward the shore. This has significant biological importance to cascading effects of food chains in coral reef ecosystems and may provide yet another key to unlocking "Darwin's Paradox".

Cyanobacteria provide soluble nitrates for the reef via nitrogen fixation.

Coral reefs also often depend on surrounding habitats, such as seagrass meadows and mangrove forests, for nutrients. Seagrass and mangroves supply dead plants and animals which are rich in nitrogen and also serve to feed fish and animals from the reef by supplying wood and vegetation. Reefs, in turn, protect mangroves and seagrass from waves and produce sediment in which the mangroves and seagrass can root.

Biodiversity

Tube sponges attracting cardinal fishes, glassfishes and wrasses

Over 4,000 species of fish inhabit coral reefs.

Coral reefs form some of the world's most productive ecosystems, providing complex and varied marine habitats that support a wide range of other organisms.Fringing reefs just below low tide level have a mutually beneficial relationship with mangrove forests at high tide level and sea grass meadows in between: the reefs protect the mangroves and seagrass from strong currents and waves that would damage them or erode the sediments in which they are rooted, while the mangroves and sea grass protect the coral from large influxes of silt, fresh water and pollutants. This level of variety in the environment benefits many coral reef animals, which, for example, may feed in the sea grass and use the reefs for protection or breeding.

Organisms can cover every square inch of a coral reef.

Reefs are home to a large variety of animals, including fish, seabirds, sponges, cnidarians (which includes some types of corals and jellyfish), worms, crustaceans (including shrimp, cleaner shrimp, spiny lobsters and crabs), mollusks (including cephalopods), echinoderms (including starfish, sea urchins and sea cucumbers), sea squirts, sea turtles and sea snakes. Aside from humans, mammals are rare on coral reefs, with visiting cetaceans such as dolphins being the main exception. A few of these varied species feed directly on corals, while others graze on algae on the reef. Reef biomass is positively related to species diversity.

The same hideouts in a reef may be regularly inhabited by different species at different times of day. Nighttime predators such as cardinalfish and squirrelfish hide during the day, while damselfish, surgeonfish, triggerfish, wrasses and parrotfish hide from eels and sharks.

Algae

Reefs are chronically at risk of algal encroachment. Overfishing and excess nutrient supply from onshore can enable algae to outcompete and kill the coral. Increased nutrient levels can be a result of sewage or chemical fertilizer runoff from nearby coastal developments. Runoff can carry nitrogen and phosphorus which promote excess algae growth. Algae can sometimes out-compete the coral for space. The algae can then smother the coral by decreasing the oxygen supply available to the reef. Decreased oxygen levels can slow down coral's calcification rates weakening the coral and leaving it more susceptible to disease and degradation. In surveys done around largely unin-habited US Pacific islands, algae inhabit a large percentage of surveyed coral locations. The algal population consists of turf algae, coralline algae, and macro algae.

Sponges

Sponges are essential for the functioning of the coral reef's ecosystem. Algae and corals in coral reefs produce organic material. This is filtered through sponges which convert this organic material into small particles which in turn are absorbed by algae and corals.

Fish

Over 4,000 species of fish inhabit coral reefs. The reasons for this diversity remain unclear. Hypotheses include the "lottery", in which the first (lucky winner) recruit to a territory is typically able to defend it against latecomers, "competition", in which adults compete for territory, and less-competitive species must be able to survive in poorer habitat, and "predation", in which population size is a function of postsettlement piscivore mortality. Healthy reefs can produce up to 35 tons of fish per square kilometer each year, but damaged reefs produce much less.

Invertebrates

Sea urchins, Dotidae and sea slugs eat seaweed. Some species of sea urchins, such as *Diadema antillarum*, can play a pivotal part in preventing algae from overrunning reefs. Nudibranchia and sea anemones eat sponges.

A number of invertebrates, collectively called "cryptofauna," inhabit the coral skeletal substrate itself, either boring into the skeletons (through the process of bioerosion) or living in pre-existing voids and crevices. Those animals boring into the rock include sponges, bivalve mollusks, and sipunculans. Those settling on the reef include many other species, particularly crustaceans and polychaete worms.

Seabirds

Coral reef systems provide important habitats for seabird species, some endangered. For example, Midway Atoll in Hawaii supports nearly three million seabirds, including two-thirds (1.5 million) of the global population of Laysan albatross, and one-third of the global population of black-footed albatross. Each seabird species has specific sites on the atoll where they nest. Altogether, 17 species of seabirds live on Midway. The short-tailed albatross is the rarest, with fewer than 2,200 surviving after excessive feather hunting in the late 19th century.

Other

Sea snakes feed exclusively on fish and their eggs. Marine birds, such as herons, gannets, pelicans and boobies, feed on reef fish. Some land-based reptiles intermittently associate with reefs, such as monitor lizards, the marine crocodile and semiaquatic snakes, such as *Laticauda colubrina*. Sea turtles, particularly hawksbill sea turtles, feed on sponges.

Schooling reef fish   Caribbean reef squid

Whitetip reef shark   Green turtle

Importance

Coral reefs deliver ecosystem services to tourism, fisheries and coastline protection. The global economic value of coral reefs has been estimated to be between US $29.8 billion and $375 billion per year. Coral reefs protect shorelines by absorbing wave energy, and many small islands would not exist without their reefs to protect them. According to the environmental group World Wide Fund for Nature, the economic cost over a 25-year period of destroying one kilometer of coral reef is somewhere between $137,000 and $1,200,000. About six million tons of fish are taken each year from coral reefs. Well-managed coral reefs have an annual yield of 15 tons of seafood on average per square kilometer. Southeast Asia's coral reef fisheries alone yield about $2.4 billion annually from seafood.

To improve the management of coastal coral reefs, another environmental group, the World Resources Institute (WRI) developed and published tools for calculating the value of coral reef-related tourism, shoreline protection and fisheries, partnering with five Caribbean countries. As of April 2011, published working papers covered St. Lucia, Tobago, Belize, and the Dominican Republic, with a paper for Jamaica in preparation. The WRI was also "making sure that the study results support improved coastal policies and management planning". The Belize study estimated the value of reef and mangrove services at $395–559 million annually.

Bermuda's coral reefs provide economic benefits to the Island worth on average $722 million per year, based on six key ecosystem services, according to Sarkis *et al* (2010).

Threats

Coral reefs are dying around the world. In particular, coral mining, agricultural and urban run-off, pollution (organic and inorganic), overfishing, blast fishing, disease, and the digging of canals and access into islands and bays are localized threats to coral ecosystems. Broader threats are sea temperature rise, sea level rise and pH changes from ocean acidification, all associated with greenhouse gas emissions. A 2014 study lists factors such as population explosion along the coast lines, overfishing, the pollution of coastal areas, global warming and invasive species among the main reasons that have put reefs in danger of extinction.

Island with fringing reef off Yap, Micronesia

A study released in April 2013 has shown that air pollution can also stunt the growth of coral reefs; researchers from Australia, Panama and the UK used coral records (between 1880 and 2000) from the western Caribbean to show the threat of factors such as coal-burning coal and volcanic eruptions. Pollutants, such as Tributyltin, a biocide released into water from in anti-fouling paint can be toxic to corals.

In 2011, researchers suggested that "extant marine invertebrates face the same synergistic effects of multiple stressors" that occurred during the end-Permian extinction, and that genera "with poorly buffered respiratory physiology and calcareous shells", such as corals, were particularly vulnerable.

Rock coral on seamounts across the ocean are under fire from bottom trawling. Reportedly up to 50% of the catch is rock coral, and the practice transforms coral structures to rubble. With it taking years to regrow, these coral communities are disappearing faster than they can sustain themselves.

Another cause for the death of coral reefs is bioerosion. Various fishes graze corals, dead or alive and change the morphology of coral reefs making them more susceptible to other physical and chemical threats. It has been generally observed that only the algae growing on dead corals is eaten and the live ones are not. However, this act still destroys the top layer of coral substrate and makes it harder for the reefs to sustain.

In El Niño-year 2010, preliminary reports show global coral bleaching reached its worst level since another El Niño year, 1998, when 16% of the world's reefs died as a result of increased water temperature. In Indonesia's Aceh province, surveys showed some 80% of bleached corals died. Scientists do not yet understand the long-term impacts of coral bleaching, but they do know that bleaching leaves corals vulnerable to disease, stunts their growth, and affects their reproduction, while severe bleaching kills them. In July, Malaysia closed several dive sites where virtually all the corals were damaged by bleaching.

To find answers for these problems, researchers study the various factors that impact reefs. The list includes the ocean's role as a carbon dioxide sink, atmospheric changes, ultraviolet light, ocean acidification, viruses, impacts of dust storms carrying agents to far-flung reefs, pollutants, algal blooms and others. Reefs are threatened well beyond coastal areas. Coral reefs with one type of zooxanthellae are more prone to bleaching than are reefs with another, more hardy, species.

General estimates show approximately 10% of the world's coral reefs are dead. About 60% of the world's reefs are at risk due to destructive, human-related activities. The threat to the health of reefs is particularly high in Southeast Asia, where 95% of reefs are at risk from local threats. By the 2030s, 90% of reefs are expected to be at risk from both human activities and climate change; by 2050, all coral reefs will be in danger.

Current research is showing that ecotourism in the Great Barrier Reef is contributing to coral disease, and that chemicals in sunscreens may contribute to the impact of viruses on zooxanthellae.

Some scientists, including those associated with the National Oceanic and Atmospheric Administration, posit that US coral reefs are likely to disappear within a few decades as a result of global warming.

Protection

A diversity of corals

Marine protected areas (MPAs) have become increasingly prominent for reef management. MPAs promote responsible fishery management and habitat protection. Much like national parks and wildlife refuges, and to varying degrees, MPAs restrict potentially damaging activities. MPAs encompass both social and biological objectives, including reef restoration, aesthetics, biodiversity, and economic benefits. However, there are very few MPAs that have actually made a substantial difference. Research in Indonesia, Philippines and Papua New Guinea shows that there is no significant difference between an MPA site and an unprotected site. Conflicts surrounding MPAs involve lack of participation, clashing views of the government and fisheries, effectiveness of the area, and funding. In some situations, as in the Phoenix Islands Protected Area, MPAs can also provide revenue, potentially equal to the income they would have generated without controls, as Kiribati did for its Phoenix Islands.

According to the *Caribbean Coral Reefs - Status Report 1970-2012* made by the IUCN. States that; stopping overfishing especially key fishes to coral reef like parrotfish, coastal zone management which reduce human pressure on reef, (for example restricting the coastal settlement, development and tourism in coastal reef) and controlling pollution specially sewage wastage, may not only reduce coral declining but also reverse it and may let to coral reef more adaptable to changes relates to climate and acidification. The report shows that healthier reef in the Caribbean are those with large population of parrotfish in countries which protect these key fishes and sea urchins, banning fish trap and Spearfishing creating "resilient reefs".

To help combat ocean acidification, some laws are in place to reduce greenhouse gases such as carbon dioxide. The Clean Water Act puts pressure on state government agencies to monitor and limit runoff of pollutants that can cause ocean acidification. Stormwater surge preventions are also in place, as well as coastal buffers between agricultural land and the coastline. This act also ensures that delicate watershed ecosystems are intact, such as wetlands. The Clean Water Act is funded by the federal government, and is monitored by various watershed groups. Many land use laws aim to reduce CO_2 emissions by limiting deforestation. Deforestation causes erosion, which releases a large amount of carbon stored in the soil, which then flows into the ocean, contributing to ocean acidification. Incentives are used to reduce miles traveled by vehicles, which reduces the carbon emissions into the atmosphere, thereby reducing the amount of dissolved CO_2 in the ocean. State and federal governments also control coastal erosion, which releases stored carbon in the soil into the ocean, increasing ocean acidification. High-end satellite technology is increasingly being employed to monitor coral reef conditions.

Biosphere reserve, marine park, national monument and world heritage status can protect reefs. For example, Belize's barrier reef, Sian Ka'an, the Galapagos islands, Great Barrier Reef, Henderson Island, Palau and Papahānaumokuākea Marine National Monument are world heritage sites.

In Australia, the Great Barrier Reef is protected by the Great Barrier Reef Marine Park Authority, and is the subject of much legislation, including a biodiversity action plan. They have compiled a Coral Reef Resilience Action Plan. This detailed action plan consists of numerous adaptive management strategies, including reducing our carbon footprint, which would ultimately reduce the amount of ocean acidification in the oceans surrounding the Great Barrier Reef. An extensive public awareness plan is also in place to provide education on the "rainforests of the sea" and how people can reduce carbon emissions, thereby reducing ocean acidification.

Inhabitants of Ahus Island, Manus Province, Papua New Guinea, have followed a generations-old practice of restricting fishing in six areas of their reef lagoon. Their cultural traditions allow line fishing, but no net or spear fishing. The result is both the biomass and individual fish sizes are significantly larger than in places where fishing is unrestricted.

Restoration

Coral aquaculture, also known as coral farming or coral gardening, is showing promise as a potentially effective tool for restoring coral reefs, which have been declining around the world. The process bypasses the early growth stages of corals when they are most at risk of dying. Coral seeds are grown in nurseries, then replanted on the reef. Coral is farmed by coral farmers who live locally to the reefs and farm for reef conservation or for income.

Coral fragments growing on nontoxic concrete

Efforts to expand the size and number of coral reefs generally involve supplying substrate to allow more corals to find a home. Substrate materials include discarded vehicle tires, scuttled ships, subway cars, and formed concrete, such as reef balls. Reefs also grow unaided on marine structures such as oil rigs. In large restoration projects, propagated hermatypic coral on substrate can be secured with metal pins, superglue or milliput. Needle and thread can also attach A-hermatype coral to substrate.

A substrate for growing corals referred to as Biorock is produced by running low voltage electrical currents through seawater to crystallize dissolved minerals onto steel structures. The resultant white carbonate (aragonite) is the same mineral that makes up natural coral reefs. Corals rapidly colonize and grow at accelerated rates on these coated structures. The electrical currents also accelerate formation and growth of both chemical limestone rock and the skeletons of corals and other shell-bearing organisms. The vicinity of the anode and cathode provides a high-pH environment which inhibits the growth of competitive filamentous and fleshy algae. The increased growth rates fully depend on the accretion activity.

During accretion, the settled corals display an increased growth rate, size and density, but after the process is complete, growth rate and density return to levels comparable to natural growth, and are about the same size or slightly smaller.

One case study with coral reef restoration was conducted on the island of Oahu in Hawaii. The University of Hawaii has come up with a Coral Reef Assessment and Monitoring Program to help relocate and restore coral reefs in Hawaii. A boat channel on the island of Oahu to the Hawaii Institute of Marine Biology was overcrowded with coral reefs. Also, many areas of coral reef patches in the channel had been damaged from past dredging in the channel. Dredging covers the existing corals with sand, and their larvae cannot build and thrive on sand; they can only build on to existing reefs. Because of this, the University of Hawaii decided to relocate some of the coral reef to a different transplant site. They transplanted them with the help of the United States Army divers, to a relocation site relatively close to the channel. They observed very little, if any, damage occurred to any of the colonies while they were being transported, and no mortality of coral reefs has been observed on the new transplant site, but they will be continuing to monitor the new transplant site to see how potential environmental impacts (i.e. ocean acidification) will harm the overall reef mortality rate. While trying to attach the coral to the new transplant site, they found the coral placed on hard rock is growing considerably well, and coral was even growing on the wires that attached the transplant corals to the transplant site. This gives new hope to future research on coral reef transplant sites. As a result of this coral restoration project, no environmental effects were seen from the transplantation process, no recreational activities were decreased, and no scenic areas were affected by the project. This is a great example that coral transplantation and restoration can work and thrive under the right conditions, which means there may be hope for other damaged coral reefs.

Another possibility for coral restoration is gene therapy. Through infecting coral with genetically modified bacteria, it may be possible to grow corals that are more resistant to climate change and other threats.

Reefs in the Past

Ancient coral reefs

Throughout Earth history, from a few thousand years after hard skeletons were developed by marine organisms, there were almost always reefs. The times of maximum development were in the Middle Cambrian (513–501 Ma), Devonian (416–359 Ma) and Carboniferous (359–299 Ma), owing to order Rugosa extinct corals, and Late Cretaceous (100–66 Ma) and all Neogene (23 Ma–present), owing to order Scleractinia corals.

Not all reefs in the past were formed by corals: those in the Early Cambrian (542–513 Ma) resulted from calcareous algae and archaeocyathids (small animals with conical shape, probably related to sponges) and in the Late Cretaceous (100–66 Ma), when there also existed reefs formed by a group of bivalves called rudists; one of the valves formed the main conical structure and the other, much smaller valve acted as a cap.

Measurements of the oxygen isotopic composition of the aragonitic skeleton of coral reefs, such as Porites, can indicate changes in the sea surface temperature and sea surface salinity conditions of the ocean during the growth of the coral. This technique is often used by climate scientists to infer the paleoclimate of a region.

Coral polyps inside a colony

Most corals seen on a coral reef are colonial, consisting of many polyps growing together to form a colony. 'Coral' is a term that encompasses a diverse array of these anemone-like animals, the most common of these being stony corals and soft corals.

Stony coral or true coral is an organism in the order Scleractinia. Organisms in this order get their name from their skeletons, which are composed of hardened calcium carbonate which can cause the coral to feel like stone. While a coral is alive, the skeleton is covered in a soft layer of living material, but after corals die, their hardened skeletons are clearly visible. Organisms in this order can be divided into two groups: colonial and solitary. Colonial stony coral forms colonies which develop into the fantastic forms many people associate with coral reefs. Solitary stony corals do not live together in colonies, and many of them are also free- floating.

In the case of a colony of stony coral, the hard skeleton is created by numerous individuals known as polyps, which work together to build the skeleton. Corals can grow asexually by budding, a process which splits the polyps into copies of themselves, and colonies can also grow by fusing with neighboring colonies. Stony coral is also capable of sexual reproduction, which is usually accomplished by releasing eggs and sperm into the ocean, where gametes can form when eggs and sperm come into contact with each other. In the case of stony coral which grows into colonies, the gametes can start new colonies.

Stony corals can also be divided into zooxanthellate and non-zooxanthellate corals. Zooxanthellate corals form symbiotic relationships with algae which live inside the coral skeleton, providing energy for the colony. Non-zooxanthellate corals, as you might imagine, do not rely on algae for food. In both cases, the polyps also supply their own food, using specialized structures known as sweeper tentacles to grab prey as it drifts by on the current.

A number of basic shapes of stony coral can be observed in the ocean, including branching corals, pillar corals, table corals, elkhorn corals, encrusting corals, massive corals, massive corals, and foliase corals, which form interconnected whorls and plates of material. All stony coral species adhere to a rocky or hard substrate, and once a coral is established, it can be extremely difficult to dislodge.

Stony corals, such as brain corals, belonging to the Order Scleractinia, secrete an external skeleton of calcium carbonate (limestone). Brain coral is a common name given to corals in the family

Faviidae so called due to their generally spheroid shape and grooved surface which resembles an animal brain. Each head of coral is formed by a colony of genetically identical polyps which secrete a hard skeleton of calcium carbonate; this makes them important coral reef builders like other stony corals in the order Scleractinia.

Brain corals are found in shallow warm-water coral reefs in all the world's oceans. They are part of the phylum Cnidaria, in a class called Anthozoa or "sea flowers." The life span of the largest brain corals is 900 years. Colonies can grow as large as 6 or more feet (1.8 m) high. Brain corals extend their tentacles to catch food at night. During the day, the brain corals use their tentacles for protection by wrapping them over the grooves on their surface. The surface is hard and offers good protection against fish or hurricanes. Branching corals, such as staghorn corals, grow more rapidly, but those are more vulnerable to storm damage.

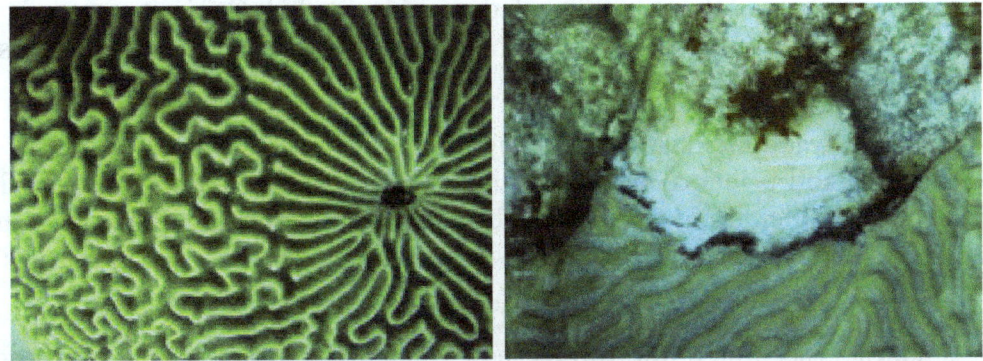

Live brain coral (left) and white skeleton of brain coral revealed after damage by black band disease.

Like other genera of corals, brain corals feed on small drifting animals and also receive nutrients provided by the algae which live within their tissues. The behavior of one of the most common genera, Favia, is semi-aggressive; it will sting other corals with its extended sweeper tentacles during the night.The genus and species has not been defined through the scientific classification segment.

Soft corals, belonging to the Order Alcyonacea, do not have large external skeletons, although most have small internal spicules of calcium carbonate instead. The gorgonians (e.g. sea fans, sea rods and sea plumes), which are most common in the Caribbean, also gain support from internal proteinaceous rods of gorgonin.

Soft corals, typified by their internal fleshy skeletons, are the most appropriate varieties of stinging animals for the marine aquarist graduating from fish to invertebrate to full-blown- reef enthusiast. Many of these are tolerant toward aquarium conditions, relatively inexpensive, and more easily cared for than the small or large polyped true or stony corals.

Beyond the above considerations is one that should be important to every conscientious aquarist; the removal of soft corals from the worlds reefs is less destructive to the environment than chipping away, otherwise removing calcareous corals. Their recruitment (growth and replacement) rates are far greater, they're not principal prey species, and little used as habitat by other reef creatures.

Classification: Taxonomy, Relation with other Groups

The soft corals are members of the stinging-celled animals, Phylum Cnidaria, formerly Coelenterata; a group that includes the anemones, jellyfishes, hydroids, sea-pens, the true corals and other coral-named groups.

Cnidarians are tissue-grade life characterized by having just two germ layers (ecto- and endoderm), stinging-cells, and principal radial symmetry. Other salient characteristics; they have a single body cavity (the coelenteron) that is sac-like, with one opening that serves as both an mouth and an anus; lack a central nervous system (have simple nerve nets), no head or gas exchange, excretory or circulatory systems.

The phylum Cnidaria is separated into three Classes roughly by the principal form (bell- shaped free-living medusoid, or attached polypoid) they take as life stages.

- Class Scyphozoa, the jellyfishes, are mainly medusoids.

- Class Hydrozoa, the hydroids and hydromedusae, display alternation of generations with asexual benthic polyps alternating with sexual planktonic medusae.

- Class Anthozoa, sea anemones, corals, sea pens; have no medusoid stage. They are further subdivided into two subclasses.

  — Subclass Hexacorallia (=Zoantharia), sea anemones and true corals (Order Scleractinia), have tentacles and mesenteries (internal body divisions) in multiples of six ("hex"), 0,1,2 siphonoglyphs (slot-like mouth/anus openings).

  — Subclass Octocorallia (=Alcyonaria), the octocorals called soft, blue, organ-pipe corals, sea fans, sea pansies. Colonial polypoids, with eight-numbered mesenteries and hollow tentacles (pinnately, i.e. side-branched like a bird's feather), one siphonoglyph. A few orders of note to aquarists:

    o Order Gorgonacea, sea fans, sea whips. Non-living central structure with living "rind" covering.

    o Order Pennatulacea, sea pens, sea pansies.

    o Order Heliporacea, blue "coral".

    o Order Stolonifera, red organ-pipe "coral" (Tubipora).

The Order Alcyonacea, the soft corals, are made up of either encrusting or erect colonies, mostly fleshy and flexible with a bizarre assortment of internal structural elements called sclerites, rendering shape and structure.

Briefly, we can see that it is not only the lack of external hard, stony, calcareous skeletons that the "true" corals (Order Scleractinia, Subclass Hexacorallia) from the soft fleshy or leathery corals, the Alcyonacea and their relatives; but major elements of body plan and symmetry (tentacular and body segmentation number, mouth-anus openings).

Natural Range

Soft corals are found worldwide, more in tropical than temperate reefs, mainly in mid-depths of 5-30 meters. Abundance on reefs in the Indo-Pacific and Red Sea is often conspicuous compared with stands of colonies in the Caribbean, Hawaii and elsewhere.

Principal Forms in the Hobby

First the general disclaimer regarding classifying difficult-to-discern life forms. The alcyonaceans that we call soft corals are really told apart only by microscopic examination of the aforementioned calcareous particles (sclerites). Therefore we will detail them more generically, as in by family and most commonly available genera.

Family Nephtheidae; Carnation, Tree, Colt Soft Corals, the genera Capnella, Dendronephthya, Nephthea, Scleronephthya, Cladiella, Lemnalia and more are known to all divers and other appreciators of tropical marine environments. They are the gorgeous warm colored (red, pink, yellow) and purplish cotton-candy looking creatures attached to reefs. Know that these colorful, branched-tree animals inflate and shrink regularly in rhythm with feeding and metabolite flushing, and that many species are difficult to care for.

Nephthea sp. in captivity and in Australia's Great Barrier Reef.

Family Alcyoniidae, the mushroom, ridged and lobed leather or toadstool corals, Lobophytum, Alcyonium, Cladiella, Sarcophyton, Sinularia With their almost unreal rubbery appearance with tentacles retracted, hardly recognizable as living, yet alone as stinging-celled animals. Mainly yellow, brown to grayish in overall uniform color. Amongst all the animals called corals, the leathers are the toughest for aquarium use; getting by on higher nutrient levels/lower water quality, and less stringent turbulence and lighting conditions.

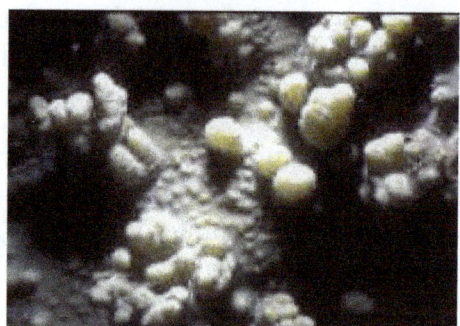

A Sinularia in Nuka Hiva, Marquesas

Family Xeniidae, the pulsing and not Xenia, Stereosoma, Anthelia, Efflatournaria with their very fine wafting colonies of long feathery tentacled polyps pulsing and waving rhythmically. White, brown to bluish in coloration.

A Xenia sp. in Australia and Anthelia glauca in the Red Sea

Family Nidaliidae, some of these soft corals are superficially very similar to gorgonians (sea fans). Though quite common in the wild (Indo-Pacific) the gorgonian-like ones (genera Chironephthya, Siphonogorgia) are almost impossible to keep in captivity. Lacking zooxanthellae Nidaliids must be fed on suspensions of small zooplankton. Other genera that more resemble Nephtheids (Agaricoidea, Nidalla, Pieterfaurea) are of use to aquarists, though not as easily kept as the more hardy members of that family.

A Siphonogorgia with retracted polyps in a typical setting off Heron Island, Australia's Great Barrier Reef

Coral Bleaching

Bleaching, or the paling of zooxanthellate invertebrates, occurs when (i) the densities of zooxanthellae decline and / or (ii) the concentration of photosynthetic pigments within the zooxanthellae fall (Kleppel et al. 1989). Most reef-building corals normally contain around 1- 5 x 106 zooxanthellae cm-2 of live surface tissue and 2-10 pg of chlorophyll a per zooxanthella. When corals bleach they commonly lose 60-90% of their zooxanthellae and each zooxanthella may lose 50-80% of its photosynthetic pigments (Glynn 1996). The pale appearance of bleached scleractinian corals and hydrocorals is due to the cnidarian's calcareous skeleton showing through the translucent tissues (that are nearly devoid of pigmented zooxanthellae). Under stress, corals may expel their zooxanthellae, which leads to a lighter or completely white appearance, hence the term "bleached". Once bleaching begins, it tends to continue even without continuing stress. If the coral colony survives the stress period, zooxanthellae often require weeks to months to return to normal density. The new residents may be of a different species. Some species of zooxanthellae and corals are more resistant to stress than other species.

If the stress-causing bleaching is not too severe and if it decreases in time, the affected corals usually regain their symbiotic algae within several weeks or a few months. If zooxanthellae loss is prolonged, i.e. if the stress continues and depleted zooxanthellae populations do not recover, the coral host eventually dies.

Three hypotheses have been advanced to explain the cellular mechanism of bleaching, and all are based on extreme sea temperatures as one of the causative factors. High temperature and irradiance stressors have been implicated in the disruption of enzyme systems in zooxanthellae that offer protection against oxygen toxicity. Photosynthesis pathways in zooxanthallae are impaired at temperatures above 30 degrees C, this effect could activate the disassociation of coral / algal symbiosis. Low- or high-temperature shocks results in zooxanthellae low as a result of cell adhesion dysfunction. This involves the detachment of cnidarian endodermal cells with their zooxanthellae and the eventual expulsion of both cell types.

It has been hypothesized that bleaching is an adaptive mechanism which allows the coral to be repopulated with a different type of zooxanthellae, possibly conferring greater stress resistance. Different strains of zooxanthellae exist both between and within different species of coral hosts, and the different strains of algae show varied physiological responses to both temperature and irradiance exposure. The coral / algal association may have the scope to adapt within a coral's lifetime. Such adaptations could be either genetic or phenotypic.

Unbleached (left) and bleached (right) coral

Causes of Coral Bleaching

As coral reef bleaching is a general response to stress, it can be induced by a variety of factors, alone or in combination. It is therefore difficult to unequivocally identify the causes for bleaching events. The following stressors have been implicated in coral reef bleaching events.

- Temperature

  Coral species live within a relatively narrow temperature margin, and anomalously low and high sea temperatures can induce coral bleaching. Bleaching events occur during sudden temperature drops accompanying intense upwelling episodes, (-3 degrees C to −5 degrees C for 5-10 days), seasonal cold-air outbreaks. Bleaching is much more frequently reported from elevated se water temperature. A small positive anomaly of 1-2 degrees C for 5-10 weeks during the summer season will usually induce bleaching.

- Solar Irradiance

  Bleaching during the summer months, during seasonal temperature and irradiance maxima often occurs disproportionately in shallow-living corals and on the exposed summits of colonies. Solar radiation has been suspected to play a role in coral bleaching. Both photosyntheticaly active radiation (PAR, 400-700nm) and ultraviolet radiation (UVR, 280-400nm) have been implicated in bleaching.

- Subaerial Exposure

  Sudden exposure of reef flat corals to the atmosphere during events such as extreme low tides, ENSO-related sea level drops or tectonic uplift can potentially induce bleaching. The consequent exposure to high or low temperatures, increased solar radiation, desiccation, and sea water dilution by heavy rains could all play a role in zooxanthellae loss, but could also very well lead to coral death.

- Sedimentation

  Relatively few instances of coral bleaching have been linked solely to sediment. It is possible, but has not been demonstrated, that sediment loading could make zooxanthellate species more likely to bleach.

- Fresh Water Dilution

  Rapid dilution of reef waters from storm-generated precipitation and runoff has been demonstrated to cause coral reef bleaching. Generally, such bleaching events are rare and confined to relatively small, nearshore areas.

- Inorganic Nutrients

  Rather than causing coral reef bleaching, an increase in ambient elemental nutrient concentrations (e.g. ammonia and nitrate) actually increases zooxanthellae densities 2-3 times. Although eutrophication is not directly involved in zooxanthellae loss, it could cause secondary adverse affects such as lowering of coral resistance and greater susceptibility to diseases.

- Xenobiotics

  Zooxanthellae loss occurs during exposure of coral to elevated concentrations of various chemical contaminants, such as Cu, herbicides and oil. Because high concentrations of xenobiotics are required to induce zooxanthellae loss, bleaching from such sources is usually extremely localized and / or transitory.

- Epizootics

Pathogen induced bleaching is different from other sorts of bleaching. Most coral diseases cause patchy or whole colony death and sloughing of soft tissues, resulting in a white skeleton. A few pathogens have been identified the cause translucent white tissues, a protozoan.

Factors that influence the outcome of a bleaching event include stress-resistance which reduces bleaching, tolerance to the absence of zooxanthellae, and how quickly new coral grows to replace the dead. Due to the patchy nature of bleaching, local climatic conditions such as shade or a stream of cooler water can reduce bleaching incidence. Coral and zooxanthellae health and genetics also influence bleaching.

Spatial and Temporal range of Coral Reef Bleaching

Mass coral moralities in coral reef ecosystems have been reported in all major reef provinces since the 1870s. The frequency and scale of bleaching disturbances has increased dramatically since the late 70's. This is possibly due to more observers and a greater interest in reporting in recent years. More than 60 coral reef bleaching events out of 105 mass coral moralities were reported between 1979-1990, compared with only three bleaching events among 63 mass coral moralities recorded during the preceding 103 years.

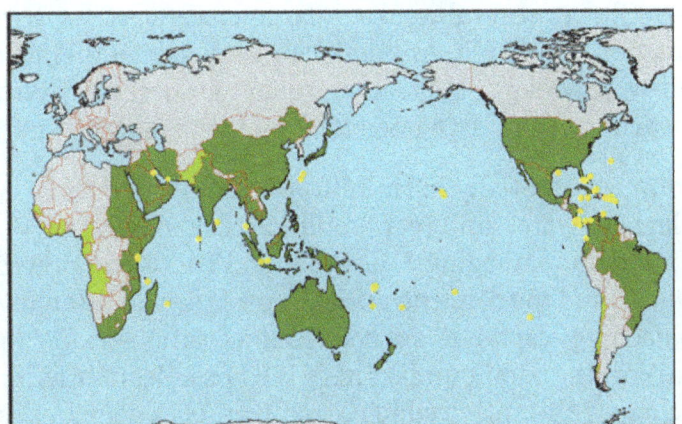

Regions where major coral reef bleaching events have taken place during the past 15 years.Yellow spots indicate major bleaching events.

Nearly all of the world's major coral reef regions (Caribbean/ western Atlantic, eastern Pacific, central and western Pacific, Indian Ocean, Arabian Gulf, Red Sea) experienced some degree of coral bleaching and mortality during the 1980s.

Prior to the 1980s, most mass coral moralities were related to non-thermal disturbances such as storms, aerial exposures during extreme low tides, and Acanthaster outbreaks. Coral bleaching accompanied some of the mortality events prior to the 1980s during periods of elevated sea water temperature, but these disturbances were geographically isolated and restricted to particular reefs zones. In contrast, many of the coral bleaching events observed in the 1980s occurred over large geographic regions and at all depths. Most of the coral reef bleaching events of the 1980s occurred during years of large-scale ENSO activity.

Global Change and Reef Bleaching

Of the causing stressors of coral reef bleaching, many are related to local environmental degradation and reef overexploitation. Of the stressors mentioned above, only sea water temperature and solar irradiance have possible global factors driving changes and extremes. Global warming, along with ENSO events, change sea water temperatures. Ozone depletion increases the amount of UVR reaching the Earth's surface, and possibly causing coral bleaching events.

Increased sea temperatures and solar radiation (especially UV radiation), either separately or in combination, have received consideration as plausible large-scale stressors. In most instances, wherever coral reef bleaching was reported, it occurred during the summer season or near the end of a protracted warming period.

Coral bleaching was reported to have occurred during periods of low wind velocity, clear skies, calm seas and low turbidity, when conditions favor localized heating and high penetration of short wave length (UV) radiation. Also less oxygen is held by water at higher temperatures. Potentially stressful high sea temperatures and UV radiation flux could conceivably cause coral reef bleaching on a global scale with suspected greenhouse warming and the thinning of the ozone layer.

As reef building corals live near their upper thermal tolerance limits, small increases in sea temperature (.5 –1.5 degrees C) over several weeks or large increases (3-4 degrees C) over a few days will lead to coral dysfunction and death. Anomalously high sea temperatures have often been reported in the Caribbean-wide series of bleaching events that occurred during 1986-88, leading to hypothesis that global warming was having an effect on the coral reefs in this region.

Solar ultraviolet radiation is potentially harmful to reef corals and their symbiotic. UV radiation can readily penetrate clear sea water, and reef –building corals contain UV- absorbing compounds capable of blocking potentially damaging UV radiation. These compounds are produced in response to ambient UV levels and the concentration in corals is usually an inverse function of depth, but it is not known if bleaching responses are related to variations in UV flux that exceed the protective capacity of UV-absorbing compounds. There is a possible interaction between temperature and UV, with temperature significantly reducing zooxanthelae densities and also the concentration of UV absorbing compounds in a reef zooanthid, thus potentially increasing the exposure of the symbionts to the direct effects of UV radiation.

If a global warming trend impacts on shallow tropical and subtropical seas, we may expect an increase in the frequency, severity and scale of coral reef bleaching. Coral mortality could exceed 95% regionally with species extirpation and extinctions. A conservative temperature increase of 1-2 degrees C would cause regions between 20-30 degrees N to experience sustained warming that falls within the lethal limits of most reef-building coral species. In conjunction with sea temperature rise would be a sea level rise, and it has been suggested that sea level rise would suppress coral growth or kill many corals through drowning or lower light levels. Some coral populations and their endosymbiotic zooxanthellae may be able to adapt to the extreme conditions predicted during global climate change. Refuges in benign habitats, such as deep, sunlit reef substrates, oceanic shoals and relatively high latitude locations, might exist, but widespread coral mortality and reef decline would be expected in shallow reef zones in most low latitude. Even if significant sea warming and elevated irradiance levels do not occur, coral reef degradation from anthropogenic pollution and overexploitation will still continue, a result of unrelenting human population growth.

Scuba Diving

Scuba diving ("scuba" originally being an acronym for Self Contained Underwater Breathing Apparatus, although now widely considered a word in its own right) is a form of underwater diving in which a diver uses a scuba set to breathe underwater for recreation, commercial or industrial reasons.

Unlike early diving, which relied exclusively on air pumped from the surface, scuba divers carry their own source of breathing gas (usually compressed air), allowing them greater freedom than with an air line. Both surface supplied and scuba diving allow divers to stay underwater significantly longer than with breath-holding techniques as used in snorkelling and free-diving.

According to the purpose of the dive, a diver usually moves underwater by swim fins attached to his feet, but external propulsion can come from an underwater vehicle, or a sled pulled from the surface.

History:

The first commercially successful scuba sets were the Aqualung open-circuit units developed by Emile Gagnan and Jacques-Yves Cousteau, in which compressed gas (usually air) is inhaled from a tank and then exhaled into the water, and the descendants of these systems are still the most popular units today. The open circuit systems were developed after Cousteau had a number of incidents of oxygen toxicity using a rebreather system, in which exhaled air is reprocessed to remove carbon dioxide. Modern versions of rebreather systems (both semi- closed circuit and closed circuit) are still available today, and form the second main type of scuba unit, most commonly used for technical diving, such as deep diving.

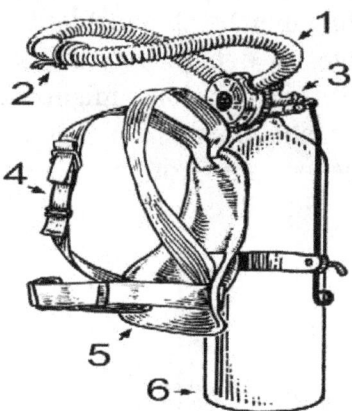

Original Aqualung SCUBA set

1. Hose

2. Mouthpiece

3. Valve

4. Harness

5. Backplate

6. Tank

Types of Diving:

Scuba diving may be performed for a number of reasons, both personal and professional. Most people begin though recreational diving, which is performed purely for enjoyment and has a number of distinct technical disciplines to increase interest underwater, such as cave diving, wreck diving, ice diving and deep diving.

Divers may be employed professionally to perform tasks underwater. Most of these commercial divers are employed to perform tasks related to the running of a business involving deep water, including civil engineering tasks such as in oil exploration, underwater welding or offshore construction. Commercial divers may also be employed to perform tasks specifically related to marine activities, such as naval diving, including the repair and inspection of boats and ships, salvage of wrecks or underwater fishing, like spear fishing.

Other specialist areas of diving include military diving, with a long history of military frogmen in various roles. They can perform roles including direct combat, infiltration behind enemy lines, placing mines or using a manned torpedo, bomb disposal or engineering operations. In civilian operations, many police forces operate police diving teams to perform search and recovery or search and rescue operations and to assist with the detection of crime which may involve bodies of water. In some cases diver rescue teams may also be part of a fire department or lifeguard unit.

Lastly, there are professional divers involved with the water itself, such as underwater photography or underwater filming divers, who set out to document the underwater world, or scientific diving, including marine biology and underwater archaeology. Reasons for diving may include:

Type of diving	Classification
aquarium maintenance in large public aquariums	commercial, scientific
boat and ship inspection, cleaning and maintenance	commercial, naval
cave diving	technical, recreational
civil engineering in harbors, water supply, and drainage systems	commercial
crude oil industry and other offshore construction and maintenance	commercial
demolition and salvage of ship wrecks	commercial, naval
diver training for reward	professional
fish farm maintenance	commercial
fishing, e.g. for abalones, crabs, lobsters, pearls, scallops, sea crayfish, sponges	commercial
frogman, manned torpedo	military
harbor clearance and maintenance	commercial, military
media diving: making television programs, etc.	professional
mine clearance and bomb disposal, disposing of unexploded ordnance	military, naval
pleasure, leisure, sport	recreational
underwater photography	professional, recreational
policing: diving to investigate or arrest unauthorized	police diving, military,
divers	naval
search and recovery diving	commercial
search and rescue diving	police
spear fishing	professional (occasionally), recreational
stealthy infiltration	military
marine biology	scientific, recreational
underwater tourism	recreational
underwater archaeology (shipwrecks; harbors, and buildings)	scientific, recreational
underwater welding	commercial

Breathing Underwater:

Water normally contains dissolved oxygen from which fish and other aquatic animals extract all their required oxygen as the water flows past their gills. Humans lack gills and do not otherwise have the capacity to breathe underwater unaided by external devices. Although the feasibility of filling and artificially ventilating the lungs with a dedicated liquid (Liquid breathing) has been established for some time, the size and complexity of the equipment allows only for medical applications with current technology.

Early diving experimenters quickly discovered it is not enough simply to supply air in order to breathe comfortably underwater. As one descends, in addition to the normal atmospheric pressure, water exerts increasing pressure on the chest and lungs—approximately 1 bar or 14.7 psi for every 33 feet or 10 meters of depth—so the pressure of the inhaled breath must almost exactly counter the surrounding or ambient pressure to inflate the lungs. It generally becomes difficult to breathe through a tube past three feet under the water.

By always providing the breathing gas at ambient pressure, modern demand valve regulators ensure the diver can inhale and exhale naturally and virtually effortlessly, regardless of depth. Because the diver's nose and eyes are covered by a diving mask; the diver cannot breathe in through the nose, except when wearing a full face diving mask. However, inhaling from a regulator's mouthpiece becomes second nature very quickly.

Open-circuit:

The most commonly used scuba set today is the "single-hose" open circuit 2-stage diving regulator, coupled to a single pressurized gas cylinder, with the first stage on the cylinder and the second stage at the mouthpiece. This arrangement differs from Emile Gagnan's and Jacques Cousteau's original 1942 "twin-hose" design, known as the Aqua-lung, in which the cylinder's pressure was reduced to ambient pressure in one or two or three stages which were all on the cylinder. The "single-hose" system has significant advantages over the original system.

In the "single-hose" two-stage design, the first stage regulator reduces the cylinder pressure of about 200 bar (3000 psi) to an intermediate level of about 10 bar (145 psi) The second stage demand valve regulator, connected via a low pressure hose to the first stage, delivers the breathing gas at the correct ambient pressure to the diver's mouth and lungs. The diver's exhaled gases are exhausted directly to the environment as waste. The first stage typically has at least one outlet delivering breathing gas at unreduced tank pressure. This is connected to the diver's pressure gauge or computer, in order to show how much breathing gas remains.

Rebreather:

Less common are closed and semi-closed rebreathers, which unlike open-circuit sets that vent off all exhaled gases, reprocess each exhaled breath for re-use by removing the carbon dioxide buildup and replacing the oxygen used by the diver.

Rebreathers release few or no gas bubbles into the water, and use much less oxygen per hour because exhaled oxygen is recovered; this has advantages for research, military, photography, and other applications. The first modern rebreather was the MK-19 that was developed at S- Tron by Ralph Osterhout that was the first electronic system. Rebreathers are more complex and more expensive than sport open-circuit scuba, and need special training and maintenance to be safely used.

Because the nitrogen in the system is kept to a minimum, decompressing is much less complicated than traditional open-circuit scuba systems and, as a result, divers can stay down longer. Because rebreathers produce very few bubbles, they do not disturb marine life or make a diver's presence known; this is useful for underwater photography, and for covert work.

Gas mixtures:

For some diving, gas mixtures other than normal atmospheric air (21% oxygen, 78% nitrogen, 1% trace gases) can be used, so long as the diver is properly trained in their use. The most commonly used mixture is Enriched Air Nitrox, which is air with extra oxygen, often with 32% or 36% oxygen, and thus less nitrogen, reducing the likelihood of decompression sickness. The reduced nitrogen may also allow for no or less decompression stop times and a shorter surface interval between dives. A common misconception is that nitrox can reduce narcosis, but research has shown that oxygen is also narcotic.

Several other common gas mixtures are in use, and all need specialized training. The increased oxygen levels in nitrox help fend off decompression sickness, however below the maximum operating depth of the mixture, the increased partial pressure of oxygen can lead to oxygen toxicity. To displace nitrogen without the increased oxygen concentration, other diluents can be used, often helium, when the resultant mixture is called trimix.

In cases of technical dives, some of the cylinders may contain different gas mixture for each phase of the dive, typically designated as Travel, Bottom, and Decompression. These different gas mixtures may be used to extend bottom time, reduce inert gas narcotic effects, and reduce decompression times.

Hazards and Dangers

Injuries due to changes in air pressure:

Divers must avoid injuries caused by changes in air pressure. The weight of the water column above the diver causes an increase in air pressure in any compressible material (wetsuit, lungs, sinus) in proportion to depth, in the same way that atmospheric air causes a pressure of 101.3kPa (14.7 pounds-force per square inch) at sea level. Pressure injuries are called barotrauma and can be quite painful, in severe cases causing a ruptured eardrum or damage to the sinuses. To avoid them, the diver equalizes the pressure in all air spaces with the surrounding water pressure when changing depth. The middle ear and sinus are equalized using one or more of several techniques, which is referred to as clearing the ears. The mask is equalized by periodically exhaling through the nose. If a drysuit is worn, it too must be equalized by inflation and deflation, similar to a buoyancy compensator.

If properly equalized, the sinus passages can stand the increased pressure of the water with no problems. However, congestion due to cold, flu or allergies may impair the ability to equalize the pressure. This may result in permanent damage to the eardrum. Although there are many dangers involved in scuba diving, divers can decrease the dangers through proper training and education. Open-water certification programs highlight diving physiology, safe diving practices, and diving hazards.

Effects of breathing high pressure gas:

- Decompression sickness.

    The diver must avoid the formation of gas bubbles in the body, called decompression sickness or 'the bends', by releasing the water pressure on the body slowly at the end of the dive and allowing gases trapped in the bloodstream to gradually break solution and leave the body, called "off-gassing." This is done by making safety stops or decompression stops and ascending slowly using dive computers or decompression tables for guidance. Decompression sickness must be treated promptly, typically in a recompression chamber. Administering enriched-oxygen breathing gas or pure oxygen to a decompression sickness stricken diver on the surface is a good form of first aid for decompression sickness, although fatality or permanent disability may still occur.

- Nitrogen narcosis

Nitrogen narcosis or inert gas narcosis is a reversible alteration in consciousness producing a state similar to alcohol intoxication in divers who breathe high pressure gas at depth. The mechanism is similar to that of nitrous oxide, or "laughing gas," administered as anesthesia. Being "narced" can impair judgment and make diving very dangerous. Narcosis starts to affect some divers at 66 feet (20 meters). At 66 feet (20 m), Narcosis manifests itself as slight giddiness. The effects increase drastically with the increase in depth. Almost all divers are able to notice the effects by 132 feet (40 meters). At these depths divers may feel euphoria, anxiety, loss of coordination and lack of concentration. At extreme depths, hallucinogenic reaction and tunnel vision can occur. Jacques Cousteau famously described it as the "rapture of the deep". Nitrogen narcosis occurs quickly and the symptoms typically disappear during the ascent, so that divers often fail to realize they were ever affected. It affects individual divers at varying depths and conditions, and can even vary from dive to dive under identical conditions. However, diving with trimix or heliox dramatically reduces the effects of inert gas narcosis.

- Oxygen toxicity

Oxygen toxicity occurs when oxygen in the body exceeds a safe "partial pressure" (PPO2). In extreme cases it affects the central nervous system and causes a seizure, which can result in the diver spitting out his regulator and drowning. Oxygen toxicity is preventable provided one never exceeds the established maximum depth of a given breathing gas. For deep dives, (generally past 180 feet / 55 meters) "hypoxic blends" containing a lower percentage of oxygen than atmospheric air are used.

- Refraction and underwater vision

Water has a higher refractive index than air; it's similar to that of the cornea of the eye. Light entering the cornea from water is hardly refracted at all, leaving only the eye's crystalline lens to focus light. This leads to very severe hypermetropia. People with severe myopia, therefore, can see better underwater without a mask than normal- sighted people.

Diving masks and diving helmets and fullface masks solve this problem by creating an air space in front of the diver's eyes. The refraction error created by the water is mostly corrected as the light travels from water to air through a flat lens, except that objects appear approximately 34% bigger and 25% closer in salt water than they actually are. Therefore total field-of-view is significantly reduced and eye-hand coordination must be adjusted.

- Controlling buoyancy underwater

To dive safely, divers need to be able to control their rate of descent and ascent in the water. Ignoring other forces such as water currents and swimming, the diver's overall buoyancy determines whether he ascends or descends. Equipment such as the diving weighting systems, diving suits (Wet, Dry & Semi-dry suits are used depending on the water temperature) and buoyancy compensators can be used to adjust the overall buoyancy. When divers want to remain at constant depth, they try to achieve neutral buoyancy. This minimizes gas consumption caused by swimming to maintain depth.

The downward force on the diver is the weight of the diver and his equipment minus the weight of the same volume of the liquid that he is immersed in; if the result is negative, that force is upwards. Diving weighting systems can be used to reduce the diver's weight and cause an ascent in an emergency. Diving suits, mostly being made of compressible materials, shrink as the diver descends, and expand as the diver ascends, creating unwanted buoyancy changes. The diver can inject air into some diving suits to counteract this effect and squeeze. Buoyancy compensators allow easy and fine adjustments in the diver's overall volume and therefore buoyancy. For open circuit divers, changes in the diver's lung volume can be used to adjust buoyancy.

- Avoiding losing body heat

Water conducts heat from the diver 25 times better than air, which can lead to hypothermia even in mild water temperatures. Symptoms of hypothermia include impaired judgment and dexterity, which can quickly become deadly in an aquatic environment. In all but the warmest waters, divers need the thermal insulation provided by wetsuits or drysuits.

In the case of a wetsuit, the suit is designed to minimize heat loss. Wetsuits are generally made of neoprene that has small gas cells, generally nitrogen, trapped in it during the manufacturing process. The poor thermal conductivity of this expanded cell neoprene means that wetsuits reduce loss of body heat by conduction to the surrounding water. The neoprene in this case acts as an insulator. The second way in which wetsuits reduce heat loss is to trap a thin layer of water between the diver's skin and the insulating suit itself. Body heat then heats the trapped water. Provided the wetsuit is reasonably well-sealed at all openings (neck, wrists, legs), this reduces water flow over the surface of the skin, reducing loss of body heat by convection, and therefore keeps the diver warm (this is the principle employed in the use of a "Semi- Dry").

In the case of a drysuit, it does exactly that: keeps a diver dry. The suit is sealed so that frigid water cannot penetrate the suit. Drysuit undergarments are often worn under a drysuit as well, and help to keep layers of air inside the suit for better thermal insulation. Some divers carry an extra gas bottle dedicated to filling the dry suit. Usually this bottle contains argon gas, because of its better insulation as compared with air.

- Avoiding skin cuts and grazes.

Diving suits also help prevent the diver's skin being damaged by rough or sharp underwater objects, marine animals or coral.

- Diving longer and deeper safely.

There are a number of techniques to increase the diver's ability to dive deeper and longer:

i. Technical diving – diving deeper than 40 metres (130 ft), using mixed gases, and/or entering overhead environments (caves or wrecks).

ii. Surface supplied diving – use of umbilical gas supply and diving helmets.

iii. Saturation diving – long-term use of underwater habitats under pressure and a gradual release of pressure over several days in a decompression chamber at the end of a dive.

Intertidal And Underwater Coral Transplantation

Damage to the coral reef systems worldwide, particularly in developing and underdeveloped countries, is particularly alarming in recent decades. Blast fishing and other destructive fishing practices have been responsible for annihilating most of nature's coral cover resulting in lower fishery productivity and marine biodiversity. Efforts in the past had been focused on developing artificial coral reefs with little or no appreciable results. In most instances, the artificial coral reefs developed simply became effective aggregating devices for important fish species artificially raising the observed fishery catch yet not really restoring the damaged environment.

Coral reef deterioration, caused by natural and/or anthropogenic factors, has been widely reported worldwide over the last two decades. Because of the slow natural recovery of reefs and their high economic potential, reconstruction efforts of various kinds and on a wide range of scales have been proposed and efforts have been made to facilitate reef redevelopment (Craik et.al., 1990; Rinkevich 1995; Guzman 1999).

Various transplantation methods have been attempted with the goal of restoring coral cover to reefs. Much of the restoration efforts to date have been focused on responding to acute episodes of damage; in particular the repair of reefs subsequent to ship groundings. Most of these efforts are located in high-energy reef front areas, using expensive methods, and requiring hundreds of hours underwater to secure dislodged coral colonies. Apparently little consideration has been given to the fact that the high-energy environments most often affected are normally dominated by stable sediment-free substrata where natural recruitment and recovery processes are most active, potentially making restoration efforts in thesehabitats unnecessary. The overall positive impact of expensive coral reef restoration projects has recently been questioned (Harriott and Fisk, 1988b; Hatcher et al., 1989; Maragos, 1992; Edwards and Clark, 1998; Birkeland, 1999; Challenger, 1999). Rather than abandoning the work, Maragos (1992) suggested that less costly methods should be developed. A recent reevaluation (Edwards and Clark, 1998), detailed the conditions where transplantation was most appropriate, and concluded that transplantation should be viewed as a tool of last resort, for use only where natural recruitment and recovery processes are failing. We concur with these sentiments.

Simple, low-tech methods of coral transplantation have been investigated for restoring coral cover to damaged lower-energy reefs, using unattached coral fragments to mimic and accelerate asexual fragment-driven reef recovery processes (Guzman, 1991,1993; Lindahl, 1998; Fox et al., 1999; Bowden-Kerby, 1997, 2001a,b). Transplanting corals into lower energy areas precludes the necessity of securing coral transplants, considerably lowering cost and effort. A high survival rate for unattached coral transplants has been demonstrated for such sheltered areas (Maragos, 1974; Harriot and Fisk, 1988a; Guzman, 1991,1993; Lindahl, 1998; Bowden-Kerby 2001a,b), particularly for rubble environments and for larger fragment sizes.

Transplanting corals directly onto sand has also been done successfully (Bowden-Kerby, 1997; 2001b), establishing that entirely new patch reefs can be created on barren sand-flat "deserts", providing for increased fish habitat. This patch reef creation process is modelled on the natural process of coral colonization of sand dominated reef areas (Bowden-Kerby 2001b), whereby storm currents sweep detached coral colonies into sandy back reef environments where larval recruitment is not possible, but where conditions for coral growth are ideal. The key factor in coral survival on sand is the large size of coral colonies, as small fragments always perish (Bowden-Kerby 2001b).

Coral transplantation in the past suffered from serious difficulties such as:

a) Coral transplants are swept away by tidal currents

b) Substrates used are not calcium bicarbonate-based which are not conducive to the growth of the transplants

c) Even if calcium bicarbonate substrates are used, they are often too light that the transplants are damaged by tidal currents as well.

The basic problem then was to develop a substrate that will hopefully address all these concerns.

Relying on natural recruitment is one possible approach (Edwards & Clark 1998), but several limitations have been reported. First, the rate of natural recruitment of corals is often so highly variable that the process can take up to several years, especially in species broadcasting their gametes (Wallace 1985; Gleason 1996; Connell et al. 1997). Species releasing larvae (i.e., planulae) have high settlement rates in some areas, but more often than not they often settle near parents and show only limited range of dispersal (Harrison and Wallace 1990). Substantial spatial variations with respect to recruitment have also been reported in many studies (Dunstan & Johnson 1998; Hughes et al. 1999); thus, many suitable habitats are too often bare due to the absence of natural recruitment. Further, coral recruits, settling on monitored surfaces, are often low in species diversity, which means a damaged reef may require a considerably long time to regain its original diversity. Another disadvantage is that recruits in nature usually suffer from high mortality and slow growth rates (Sato 1985). For example, in southern Taiwan, less than 5% of coral recruits reportedly survived 22 months, and some of those that did were still below 1 cm in diameter (Soong, unpublished data). Moreover, some destruction renders the bottom of the sea into an unconsolidated substrate that inhibits successful recruitment of corals.

As a second approach, the transplantation of whole coral colonies has been undertaken in area of small-scale destruction where hard substrate is still available (Maragos 1974; Hudson & Diaz 1988; Munoz-Chagin 1997). Aside from providing new colonies to the receiving sites, this procedure may also increase local recruitment as in the case when transplants are gravid (Richmond & Hunter 1990). The fixation of colonies is critical in this kind of operation, but unfortunately it too often necessitates high costs (Kaly 1995). An adequate source of live corals is also a prerequisite so that enough raw materials for transplantation can be supplied without damaging the communities of the donor sites. For obvious reasons this approach can only be realistically applied in cases of certain small-scale reconstruction.

A third approach, the transplantation of coral fragments, is also feasible (Alcala et al. 1982; Oren & Benayahu 1997; Guzman 1999). Because corals are modular organisms, small pieces of corals have the capability of growing in the same way as whole colonies (Connell 1973; Birkeland et al. 1979). The growth of corals from fragments is, in fact, an important natural process, at least in some branching species (Highsmith 1982; Wallace 1985). The natural fragments first anchor and secure themselves in the crivices, more or less by chance, then continue to attach themselves to the substrate by regeneration and extension of soft tissues and skeleton. A new coral colony may start by this asexual means, and there is evidence suggesting that fragmentation is adaptive in some species (Brazeau & Lasker 1992; Fong & Lirman 1995). In artificial transplantation, exactly how to best arrange fragments, which are of course much smaller in size than whole colonies, is critical to their survival and growth rates. Because sedimentation often causes mortality and inhibits coral growth (Yap & Gomez 1985; Nagelkerken et al. 2000), keeping fragments above the bottom can reduce the likelihood of their being covered with silt. Placing a coral nursery structure in shallow waters may facilitate coral growth on account of the higher light intensities; nevertheless, the risk of the structures being easily damaged by strong wave actions is greatly increased (Plucer- Rosari & Randall 1987).

Strategically, it is imperative to consider all options available at the particular site being considered for reconstruction. Because no single method stands out as suitable for all situations and purposes, critical evaluations must be made before human intervention is taken.

Coral reefs have long been known for their rich diversity of fish and invertebrates, but examining the diversity of highly mobile fish in the open ocean has been elusive. Fish are the most prominent mobile animals on coral reefs, and achieve a level of local diversity that is rarely found among terrestrial vertebrates. The high fish diversity is unusual in that it occurs along with high total densities of individuals and high total biomass. Another unusual feature of the high fish diversity is the large number of closely related species found on most reefs.

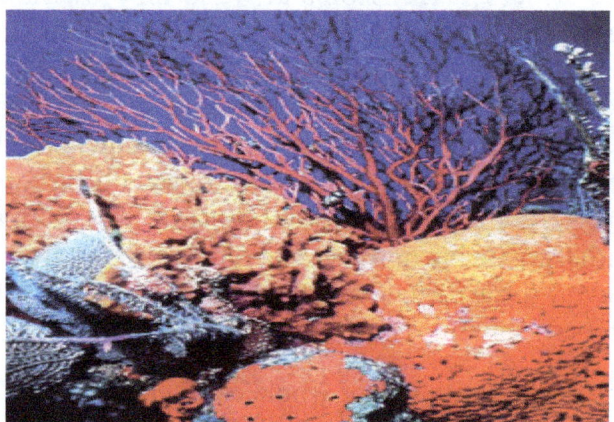

For example, in the Capricon group of reefs at the southern end of Australia's Great Barrier Reef, there are around 850 species of fish, representing 84 families and 297 genera. A number of genera have over a dozen species, including Chaetodon, Scarus, Apogon, Pomacentrus, Acanthurus and Halichoeres.

Fifty or more species commonly coexist on patch reefs only three meters in diameter, and even more species can be found coexisting within a similarly sized area at the northern end of the Great Barrier Reef, where fish diversity is even higher (Sale, 1978). equilibrium from occuring. A similar phenomenon is found in the great rift lakes of Africa (Malawi, Tanganyika, Victoria) (Lowe-McConnell, 1987).

Coral reefs are the most complex ecosystems in the seas. Fish communities reach their highest degree of diversity in these ecosystems, and differ enormously within and between reefs in the same area (William, 1991; Ormond and Roberts, 1997) and between geographic regions (Briggs, 1974, 1996). The relative roles of local and regional processes in explaining community diversity in marine systems, as well as in terrestrial systems, have been hotly debated and several, most often contradictory, explanations have been proposed (Strong et.al., 1984; Ricklefs, 1987).

The high level of diversity supported by coral reefs may be best explained as the result of various processes operating on different scales in space and time (Jackson, 1991; Kohn, 1997). At the local scale (e.g., within reef zones), the diversity observed in local fish assemblages is explained by both deterministic (interspecific competition for food and shelter; predation pressure) and stochastic (recruitments; perturbation) ecological processes (Scale, 1977, 1991; Harmelin- Vivien, 1989).

On the regional scale (e.g., Pacific vs. Atlantic, West Pacific vs. Central Pacific), the diversity of extant faunas of reef fishes is explained mainly by interactions of historical hydrodynamic and geological processes with each species' life cycle characteristics, particularly larval dispersal ability (Victor, 1991). On the global scale (e.g., tropical vs. temperate), explaining why tropical regions contain so many species has been one of problems of community ecology since the nineteenth century (Pianka, 1966; MacArthur, 1972; Stevens, 1989; Crame and Clarke, 1997), despite intensive studies in both aquatic and terrestrial environments. Until now, no convincing explanation in terms of physiology, ecology, or evolutionary processes has been offered (A. Clarke, 1996). Some arguments state that the high diversity of fishes observed on present day coral reefs is partly related to the sustained higher temperatures in the tropics over geological time, and to the more efficient use and transfer of energy permitted by long-term temperature stability. High temperature and environmental stability have influenced evolutionary processes form the molecular level to the community level of organization.

Erosion of Fish Diversity

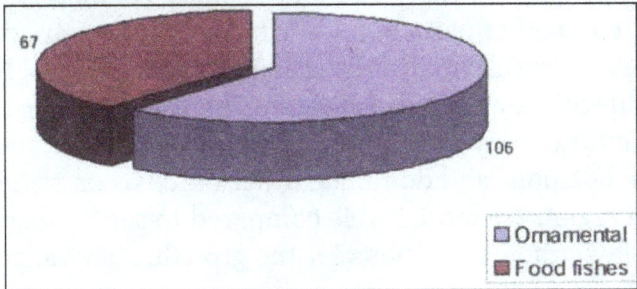

The freshwaters of India have been viewed from a single perspective: that of economic production. They are to be sources of irrigation or urban-industrial water supply or of hydel power; they are to receive sewage and industrial waste; they may produce edible fish. In this strictly utilitarian framework, there is no space to conserve the rich heritage of freshwater fish diversity of the country. All over India, freshwater fish diversity is on a decline. Many of them have been lost forever Few studies have been carried out so far regarding this aspect. They mainly identified three major forces driving extinction which are; over –harvesting, competition by newly introduced exotic fishes and pollution.

According to a workshop estimate hosted out by National Bureau Fish Genetic Resources a total of 227 Indian freshwater fishes are threatened based on the IUCN Red list Categories of 1994. The species that suffered much are Indian long fin eel (Anguilla bengalensis), the redfinned Mahseer, the catfish (Rita pervimentata), Chitala (Notoptrus chitala), smaller fishes like Indian Hatchet fish (Chela laubuca), Scarletbanded Barb (Puntis amphibious), Indian Tiger Barb (Puntis filamentous) to name a few.

Some other factors are also contributing towards this biodiversity erosion. In the irrigation canal when water is stopped in the canals, they are trapped near the gate and fished out. The nets used for the fishing often have very small mesh and so everything is caught. The shallow streams and pools, such as those at the base of waterfalls, fall victim to the easy availability of dynamite ever since quarrying and road construction began on a grand scale in the country. The shock waves of the blast destroy all fish in the vicinity. Sewage, industrial effluents, chemical fertilizers and pesticides are polluting India's freshwaters. Several carps and barbs as well as fresh water prawns are being susceptible to pollution. The drastic modification of freshwater habitats by damming streams and rivers siltation leading to reduction in their depth has also profoundly affected many fish species like the Indian shad (Hilsa ilisha), the carps (Labeo calbasu), the catfish (Bagarius bagarius) etc. Due to changed habitat, the life cycles of these species have been seriously disrupted. Moreover exotic species like Tilapia, the silver carps, the grass carps, the African catfishes proved catastrophic for native species. Its prolific breeding nature simply crowd out its native competitors. The overall deterioration of habitat has rendered many fishes susceptible to diseases. One of the most serious is epizootic ulcerative syndrome disease that brought mass mortalities and extinction of some species in Indian freshwater fishes.

Fishes-most Diverse, yet Most Neglected

Fishes are the most numerous vertebrates living on this earth and worldwide there are over 25000 species of fishes. Of this about 48% live in freshwaters that constitute just 0.01% of the earth's water. Freshwater fish diversity is unevenly distributed on this planet. The species richness is high in tropical region compared to other parts of the earth. Usually these regions are characterized by high levels of endemism. The world's major rivers like Amazon, Congo, Nile, etc. are some of the pristine rivers of the world with respect to freshwater fish diversity. I t has been estimated that the river Amazon and its tributaries may together harbour 3000 or more species of fishes. Such species-rich areas are called 'hotspots' and dominate other patterns or trends. Probably the climatic conditions of the tropical region are more stable compared to the temperate regions of the world. This could be one of the favourable conditions for the growth, survival and evolution for the species in tropical regions.

While a great deal of attention has been given to the loss of biodiversity in tropical rain forests, or in coastal areas, the diversity of and within freshwaters has been widely neglected. There is little doubt that freshwater fishes represent the most threatened set of vertebrates (Leveque, 1997). In classifying the worlds' top 25 biodiversity hotspots, vertebrate group was considered excluding fish. This is mainly because of the poorly available data wherein the author (Myer et al 2000) predicts that there could be at least 5,000 species waiting to be discovered among fish, which is more than all mammals.

Freshwater Species Population Index

Between 1970 and 1999, the Freshwater Species Population Index fell by nearly 50%, which constitutes a very rapid decline in population indices.

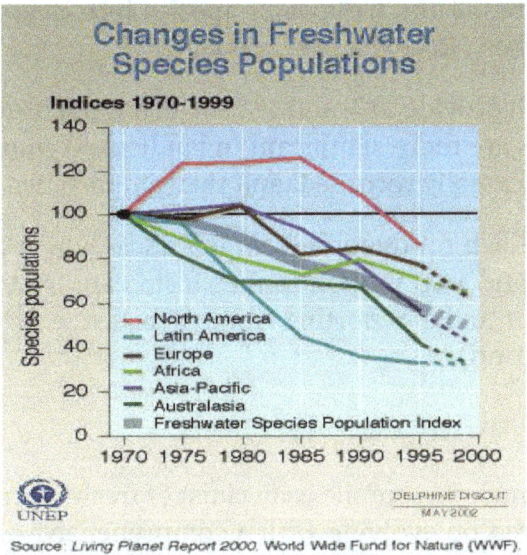

Source: *Living Planet Report 2000*, World Wide Fund for Nature (WWF).

The Freshwater Species Population Index measures the average change over time in the populations of some 194 species of freshwater birds, mammals, reptiles, amphibians and fish. The index represents the average of six regional indices, which measure freshwater species populations in Africa, Asia-Pacific, Australasia, Europe, Latin America and the Caribbean, and North America. There has been a much smaller decline over the past 30 years in the freshwater species of North America and Europe than those in the other regions. Much of the loss and degradation of freshwater ecosystems in the industrialized world took place prior to 1970.

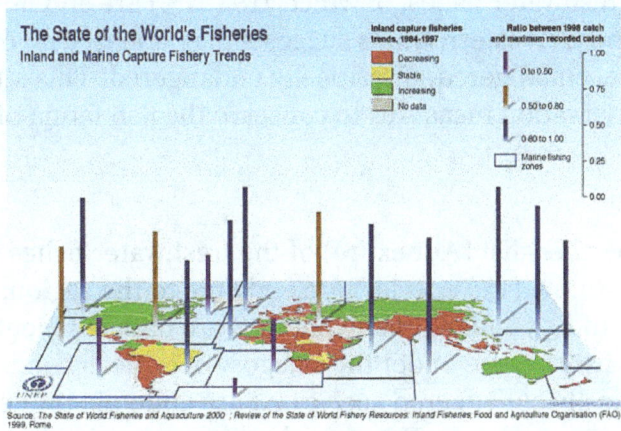

Source: *The State of World Fisheries and Aquaculture 2000 ; Review of the State of World Fishery Resources: Inland Fisheries.* Food and Agriculture Organisation (FAO). 1999. Rome.

The harvest of freshwater fish is likely to increase either through capture fisheries or aquaculture (otherwise known as 'fish farming'). In many developing countries, freshwater fish provide a significant contribution to the diets of local communities.

- The introduction of the non-native Nile Perch to Africa's Lake Victoria in 1954, combined with pollution loading and increased water turbidity resulting from agriculture and industrial development, has greatly reduced indigenous fish populations. Kenya for example, reported only 0.5% of its commercial fish catch as Nile Perch in 1976. Five years late, the proportion was 68%. Lake Victoria, the second largest lake in the world, has lost an estimated 200 different endemic species found nowhere else, while the remaining 150 are endangered. Two-thirds of the freshwater species introduced into the tropics worldwide have become established (Revenga et al., 1998)

- In Africa and Asia, fish provide 21% and 28% of all animal protein, respectively (Revenga et al., 1998). The figures are more significant in landlocked countries, where data on the fish caught are often not formally recorded, and their importance is not fully known.

- In 1999, the reported fish production form inland waters totaled 28 million tonnes, with contributions of 8.2 and 19.8 million tonnes from capture fisheries and aquaculture, respectively. With major under-reporting from subsistence fisheries, these figures could be twice as high (FAO, 2000).

Freshwater Fish Diversity of Western Ghats

Several attempts have been made to compile a checklist of freshwater fishes of the Western Ghats. These attempts mainly focused on evolving with a comprehensive checklist of freshwater fishes, which is an outcome of the patchy (may be of a river basin, a region in the Western Ghats, an administrative boundary within the Western Ghats, etc) taxonomic information available on the diversity of freshwater fishes. Daniels (2001) has listed 218 species from the Western Ghats of which 114 (52%) are endemic to Western Ghats. However, this report lacks a detailed checklist of fishes found in the Western Ghats. The subsequent checklist (Shaji et al 2001) listed 287 fishes with names of individual species. This compilation considered certain estuarine fishes that are found to ascend freshwater for longer distances. The list highlighted the presence of 67% endemic species and 18 exotic or transplanted to the region. The most recent information available is by Dahanukar et al 2004 that lists 288 freshwater fishes, of which 118 (41%) are endemic to Western Ghats. The threat status of fishes found in Western Ghats suggests that at least 41% of fish fauna is threatened by either being vulnerable, endangered, or critically endangered. This study also necessitates the implication of potent conservation measures to conserve the fish fauna of Western Ghats.

Present Scenario

Present compilation of the checklist (Annexure) of the freshwater fishes in Western Ghats region lists 318 species of which 42.8% (136 species) are endemic to the region. Of this about 27 species are critically endangered and 55 endangered while 128 are data deficient. Altogether, 39.1% (123 species) of the freshwater fishes come under the category of critically endangered, endangered and vulnerable. Of the 27 critically endangered species 24 are endemic to the region. Similarly, of the 55 endangered species, 37 are endemic. Yet 49 endemic species are data deficient. A comparison of IUCN status between endemic and non-endemic species has been made in figure, which clearly shows that the endemic species comprises more of threatened species and the non-endemic comprise more of generalist species in Western Ghats.

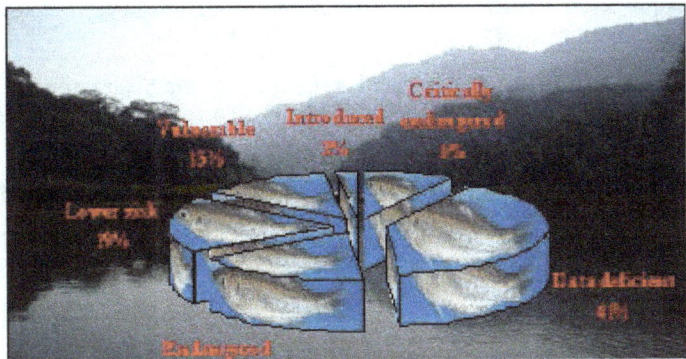

Composition with respect to IUCN status

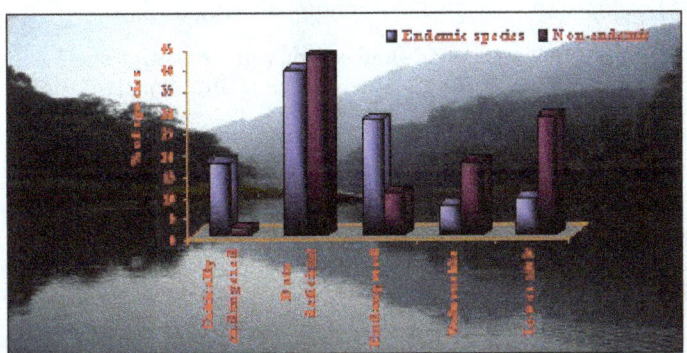

Comparison of the IUCN status between endemic and non endemic groups of fish species

References

- Fratantoni, D.; Richardson P. (2006). "The Evolution and Demise of North Brazil Current Rings". Journal of Physical Oceanography. 36 (7): 1241–1249. Bibcode:2006JPO....36.1241F. doi:10.1175/JPO2907.1

- Moyle, Peter B.; Joseph J. Cech (2004). Fishes : an introduction to ichthyology (Fifth ed.). Upper Saddle River, N.J.: Pearson/Prentice Hall. p. 556. ISBN 978-0-13-100847-2

- Cesar, H.J.S; Burke, L.; Pet-Soede, L. (2003). The Economics of Worldwide Coral Reef Degradation (PDF). The Netherlands: Cesar Environmental Economics Consulting. p. 4. Retrieved 21 September 2013

- Spalding MD, Grenfell AM (1997). "New estimates of global and regional coral reef areas". Coral Reefs. 16 (4): 225–230. doi:10.1007/s003380050078

- Gregg, M. (1989). "Scaling turbulent dissipation in the thermocline". Journal of Geophysical Research. 9686–9698. 94: 9686. Bibcode:1989JGR....94.9686G. doi:10.1029/JC094iC07p09686

- Marshall, Paul; Schuttenberg, Heidi (2006). A Reef Manager's Guide to Coral Bleaching. Townsville, Australia: Great Barrier Reef Marine Park Authority. ISBN 1-876945-40-0

- Roach, John (November 7, 2001). "Rich Coral Reefs in Nutrient-Poor Water: Paradox Explained?". National Geographic News. Retrieved April 5, 2011

- Smithers, S.G.; Woodroffe, C.D. (2000). "Microatolls as sea-level indicators on a mid-ocean atoll". Marine Geology. 168 (1–4): 61–78. doi:10.1016/S0025-3227(00)00043-8

- Taylor, J. (1992). "The energetics of breaking events in a resonantly forced internal wave field". Journal of Fluid Mechanics. 239: 309– 340. Bibcode:1992JFM...239..309T. doi:10.1017/S0022112092004427

- Crossland CJ (1983) "Dissolved nutrients in coral reef waters In DJ Barnes (Ed) Perspectives on coral reefs, pages 56–68, Australian Institute of Marine Science. ISBN 9780642895851

- McClellan, Kate; Bruno, John (2008). "Coral degradation through destructive fishing practices". Encyclopedia of Earth. Retrieved October 25, 2008

- Vajed Samiei, J.; Dab K.; Ghezellou P.; Shirvani A. (2013). "Some Scleractinian Corals (Class: Anthozoa) of Larak Island, Persian Gulf". Zootaxa. 3636 (1): 101–143. doi:10.11646/zootaxa.3636.1.5

- Helfrich, K. (1992). "Internal solitary wave breaking and run-up on a uniform slope". Journal of Fluid Mechanics. 243: 133–154. Bibcode:1992JFM...243..133H. doi:10.1017/S0022112092002660

- Osborne, Patrick L. (2000). Tropical Ecosystem and Ecological Concepts. Cambridge: Cambridge University Press. p. 464. ISBN 0-521-64523-9

- Milman, Oliver (May 30, 2017). "Scientists warn US coral reefs are on course to disappear within decades". The Guardian. Retrieved June 1, 2017

- Leichter, J.; Stewart H.; Miller S. (2003). "Episodic nutrient transport to Florida coral reefs". Limnology and Oceanography. 48 (4): 1394–1407. doi:10.4319/lo.2003.48.4.1394

- Sandstrom, H.; Elliott J. (1984). "Internal tide and solitons on the Scotian shelf: A nutrient pump at work". Journal of Geophysical Research. 89: 6415–6426. Bibcode:1984JGR....89.6415S. doi:10.1029/JC089iC04p06415

- Voris, Harold K. (1 January 1966). "Fish Eggs as the Apparent Sole Food Item for a Genus of Sea Snake, Emydocephalus (Krefft)". Ecology. 47 (1): 152. doi:10.2307/1935755

- "A biodiversity strategy for the Great Barrier Reef". Great Barrier Reef Marine Park Authority, Australian Government. Retrieved 20 September 2013

- Ferse, SCA (2010). "Poor Performance of Corals Transplanted onto Substrates of Short Durability". Restoration Ecology. 18 (4): 399–407. doi:10.1111/j.1526-100X.2010.00682.x

Human Impact on Aquatic Biodiversity

Population explosion, accompanied with urbanization and industrialization, has led to a hazardous impact on aquatic life. With the human populace indulging in overfishing, the food chain can become disturbed, and cause extinction or overpopulation of fish species. This section has been carefully written to provide an easy understanding of the varied facets of human impact on aquatic biodiversity.

Impacts of Human

Throughout history people have caused themselves problems by not understanding the consequences of their actions. Many people tend to think of fish species in isolation, but each species is part of a complex ocean ecosystem. There are interactions between fish, plankton, nutrients, water and air, if we don't understand how these interactions work, we can upset the relationships between species or between species and their habitat. So, without meaning to, we upset the balance of oceanic ecosystems.

People have been fishing since prehistoric times, for food, for profit, and for leisure, but it is only in the last few decade that there has been serious concern about overfishing, This has been partly caused by a dramatic increase in the world's population and greater international demand for food, fish meal and other marine products.

Both recreational and commercial fishing are now big business, Some commercial boats are large and are capable of staying at sea for months at a time, processing fish as they catch it, Some big boats can process 100 tonnes of fish a day. Fish yields have increased nearly fivefold over the past four decades. The current reported world total marine catch is about 90 - 100 million tonnes. The actual catch could be 30 - 50 percent higher.

At the same time, there has been an improvement in fish- catching technology, especially in deep-water fisheries. Processing, storage, transportation, and marketing systems have also improved, For example, fish finders are now used to locate fish and whole fish can now be packed in ice and sent by plane to arrive fresh or even live in overseas markets.

Human Impacts on Aquatic Biodiversity

The fish resources are no longer considered to be infinite. But at the same time, the current thinking is that fish is renewable sources. Regulations of proper inputs can make the fishery as a sustainable process, if it is tuned with the ecosystem of which it is an end product. Fish technology is very diverse, embracing aspects as varied as biology and bionomics, fish detection and location of fish stock, fish behavior.

During 1950's and 1960's witnessed the explosion of fishing technologies. Use of radars and sonar helped in detection and location of fish schools, boosting up the exploitation of the high seas and oceans. Improvement in fishing vessels, enable to reach close to the fishing grounds and improvement in gears for industrial fishing to deal with capture the bulk amount of catches further added for the exploitation, there by the over fishing eventually becomes a common practice in industrial fishing. Globally about 70% of conventionally preferred food species are subjected to overfishing.

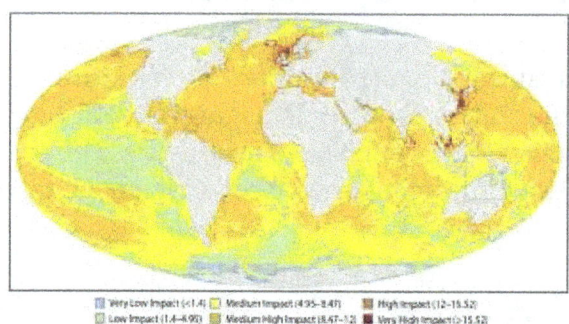

Last few decades the world wide fish catches declined, that were at peak in the Atlantic, Pacific and Mediterranean oceans where as Indian ocean was the last to be subjected to over fishing. The impact of the over fishing was felt in many ways. More over fishing to meet the supplies made the situation worse. Subsidies coming from respective Governments rise to meet out the deficits in the fishing industry and to keep employment. Resort to fishing of species situated lower in the food web and having lesser food value boomeranged in loss of food to large wild fish, causing further declining in their population. The commercial important alternative resources to wild fish [fish culture, shrimp culture or the other aquaculture practices] did not bring much improvement in the situation in the face of ever increasing fish demand resulting from ever increasing growth of human population.

Area of ocean before and after a trawler net, acting like a giant plow, scraped it.

Human activities have destroyed, disrupted or degraded a large proportion of the world's coastal, marine and freshwater ecosystems. Approximately 20% of the world's coral reefs have been destroyed. During the past 100 years, sea levels have risen 10-25 centimeters. We have destroyed more than 1/3 of the world's mangrove forests for shipping lanes.

Why is it Difficult to Protect Aquatic Biodiversity?

- Almost half of the world's people live on or near a coastal zone and 80% of ocean water pollution comes from land-based human activities.

- Increasing human population has also posed a great threat.

- Each year plastic items dumped from ships and left as litter on beaches threaten marine life.

- Rapid increasing human impacts, the invisibility of problems, citizen unawareness, and lack of legal jurisdiction hinder protection of aquatic biodiversity.

- Human ecological footprint is expanding.

- Much of the damage to oceans is not visible to most people.

- Many people incorrectly view the oceans as an inexhaustible resource.

Pollution in Freshwater Ecosystems

Fish mortality from water pollution

As with all ecosystems, the existence and operations of human society inevitably have an effect on the way of life in a freshwater community. Particularly in Western society, where a huge amount of resources are harnessed from the land to fund our lifestyle, there is a resulting effect on the ecosystems of our planet.

- Hot water is used in many industries to cool machinery. This water is removed via a discharge pipe into the river. This increase in temperature can affect the level of oxygen freely available to organisms, which, in turn affects respiration and essentially their way of life. Due to this temperature change, life in the ecosystem is affected.

- Removal of foliage next to a freshwater ecosystem allows more running water to enter its capacity. In light of this, periods of heavy rainfall can result in the water levels fluctuating wildly, which in turn can also affect the temperature of the water quite considerably not to mention all the new chemical agents that would enter the stream from this extra water.

- Recreational use of water bodies such as canoeing also has their effect. Litter from these people can sit on the surface of water and block out sunlight required by the primary producers for photosynthesis. If these primary producers way of life is affected in such a way that their population level decreases, there is a knock on effect to all those organisms who rely on these primary producers for survival.

- At a molecular level, chemicals discharged into the water, notably from industry or pesticides from farmland can affect the freshwater environment considerably. Higher concentrations of particular chemicals (perhaps toxic) mean a lower concentration of essential chemicals required by the organisms of the ecosystem. If this is the case, these organisms cannot perform respiration and function at an optimum level, thus reducing overall biomass in the ecosystem.

Freshwater pollution is the contamination of inland water (not saline) with substances that make it unfit for its natural or intended use. Pollution may be caused by faecal waste, chemicals, pesticides, petroleum, sediment, or even heated discharges. Polluted rivers and lakes are unfit for swimming or fishing; polluted water is unsafe to drink. There have been listed some 1500 substances as pollutants in freshwater ecosystems, and each of them occurs in the following types of freshwater pollutants. Some of them are:

1. Acids & Alkalis

Unpolluted deposition (or rain), in balance with atmospheric carbon dioxide, has a pH of 5.6. Almost everywhere in the world the pH of rain is lower than this. The main pollutants responsible for acid deposition (or acid rain) are sulfur dioxide (SO_2) and nitrogen oxides (NO_x). Acid deposition influences mainly the pH of freshwater.

Nitrogen and sulfuric emissions come from natural and anthropogenic sources. Natural emissions include e.g. volcano emissions, lightning, and microbial processes. Power stations and industrial plants, like the mining and smelting of high-sulfur ores and the combustion of fossil fuels, emit the largest quantities of sulfur and nitrogen oxides and other acidic compounds. These compounds mix with water vapor at unusual proportions to cause acid deposition with a pH of 4.2 to 4.7. That is 10 or more times the acidity of natural deposition.

The acidification of freshwater in an area is dependent on the quantity of calcium carbonate (limestone) in the soil. Limestone can buffer (neutralize) the acidification of freshwater. The effects of acid deposition are much greater on lakes with little buffering capacity. Much of the damage to aquatic life in sensitive areas with this little buffering capacity is a result of acid shock'. This is caused by the sudden runoff of large amounts of highly acidic water and aluminium ions into lakes and streams, when snow melts in the spring or after unusually heavy rains.

Effects on Aquatic Life

Most freshwater lakes, streams, and ponds have a natural pH in the range of 6 to 8. Acid deposition has many harmful ecological effects when the pH of most aquatic systems falls below 6 and especially below 5.

Here are some effects of increased acidity on aquatic systems:

- As the pH approaches 5, non-desirable species of plankton and mosses may begin to invade, and populations of fish such as smallmouth bass disappear.

- Below a pH of 5, fish populations begin to disappear, the bottom is covered with undecayed material, and mosses may dominate nearshore areas.

- Below a pH of 4.5, the water is essentially devoid of fish.

- Aluminium ions (Al3+) attached to minerals in nearby soil can be released into lakes, where they can kill many kinds of fish by stimulating excessive mucus formation. This asphyxiates the fish by clogging their gills. It can also cause chronic stress that may not kill individual fish, but leads to lower body weight and smaller size and makes fish less able to compete for food and habitat.

- The most serious chronic effect of increased acidity in surface waters appears to be interference with the fish' reproductive cycle. Calcium levels in the female fish may be lowered to the point where she cannot produce eggs or the eggs fail to pass from the ovaries or if fertilized, the eggs and/or larvae develop abnormally (EPA, 1980).

Extreme pH can kill adult fish and invertebrate life directly and can also damage developing juvenile fish. It will strip a fish of its slime coat and high pH level chaps the skin of fish because of its alkalinity.

When the pH of freshwater becomes highly alkaline (e.g. 9.6), the effects on fish may include: death, damage to outer surfaces like gills, eyes, and skin and an inability to dispose of metabolic wastes. High pH may also increase the toxicity of other substances. For example, the toxicity of ammonia is ten times more severe at a pH of 8 than it is at pH 7. It is directly toxic to aquatic life when it appears in alkaline conditions. Low concentrations of ammonia are generally permitted for discharge.

2. Anions

The term cyanide refers to a singularly charged anion consisting of one carbon atom and one nitrogen atom joined with a triple bond, CN-. The most toxic form of cyanide is free cyanide, which includes the cyanide anion itself and hydrogen cyanide, HCN, either in a gaseous or aqueous state. One teaspoon of a 2% cyanide solution can kill a person.

At a pH of 9.3 - 9.5, CN- and HCN are in equilibrium, with equal amounts of each present. At a pH of 11, over 99% of the cyanide remains in solution as CN-, while at pH 7, over 99% of the cyanide will exist as HCN. Although HCN is highly soluble in water, its solubility decreases with increased temperature and under highly saline conditions. Both HCN gas and liquid are colorless and have the odor of bitter almonds, although not all individuals can detect the odor.

Cyanide is frequently used in a mining technology called cyanide heap leaching. It is a cheap way to extract gold from its ore. Goldminers spray a cyanide solution (which reacts with gold) into huge open-air piles of crushed ore. They then collect the solution in leach beds and overflow ponds, recirculate it a number of times, and extract gold from it.

A problem with this technology is that cyanide is extremely toxic to birds and mammals drawn to cyanide solution collection ponds as a source of water. These ponds also can leak or overflow, posing threats to underground drinking water supplies and wildlife in lakes and streams. Because cyanide breaks down heavy metals, it can form complexes with other metals or chemicals, which can be as toxic as cyanide itself. Especially fish and aquatic invertebrates are particularly sensitive to cyanide exposure. It blocks the absorption of oxygen by cells and causes the species to suffocate. Aquatic lifes are killed by cyanide concentrations in the microgram per liter (part per billion) range, whereas bird and mammal deaths result from cyanide concentrations in the milligram per liter (part per million) range. Concentrations of free cyanide in the aquatic environment ranging from 5.0 to 7.2 micrograms per liter reduce swimming performance and inhibit reproduction in many species of fish. Other adverse effects include delayed mortality, pathology, susceptibility to predation, disrupted respiration, osmoregulatory disturbances and altered growth patterns.

Concentrations of 20 to 76 micrograms per liter free cyanide cause the death of many species, and concentrations in over 200 micrograms per liter are rapidly toxic to most species of fish. Invertebrates experience adverse nonlethal effects at 18 to 43 micrograms per liter free cyanide, and lethal effects at 30 to 100 micrograms per liter. Chronic cyanide exposure may affect reproduction, physiology, and levels of activity of many fish species, and may render the fishery resource non-viable. The sensitivity of aquatic organisms to cyanide is highly species specific, and is also affected by water pH, temperature and oxygen content, as well as the life stage and condition of the organism.

Algae and macrophytes can tolerate much higher environmental concentrations of free cyanide than fish and invertebrates, and do not exhibit adverse effects until 160 micrograms per liter or more. Aquatic plants are unaffected by cyanide at concentrations that are lethal to most species of freshwater fish and invertebrates.

Under aerobic conditions, microbial activity can degrade cyanide to ammonia, which then oxidizes to nitrate. This process has been shown effective with cyanide concentrations of up to 200 parts per million. Although biodegradation also occurs under anaerobic conditions, cyanide concentrations greater than 2 parts per million are toxic to these microorganisms.

3. Detergents

Detergents are organic compounds, which have both polar and non-polar characteristics. They tend to exist at phase boundaries, where they are associated with both polar and non-polar media. Detergents are of three types: anionic, cationic, and non-ionic. Anionic and cationic have permanent negative or positive charges, attached to non-polar (hydrophobic) C-C chains. Non-ionic detergents have no such permanent charge; instead, they have a number of atoms which are weakly electropositive and electronegative. This is due to the electron-attracting power of oxygen atoms.

There are two kinds of detergents with different characteristics: phosphate detergents and surfactant detergents. Detergents that contain phosphates are highly caustic, and surfactant detergents are very toxic. The differences are that surfactant detergents are used to enhance the wetting, foaming, dispersing and emulsifying properties of detergents. Phosphate detergents are used in detergents to soften hard water and help suspend dirt in water.

Detergents are very widely used in both industrial and domestic premises like soaps and detergents to wash vehicles. The major entry point into water is via sewage works into surface water. They are also used in pesticide formulations and for dispersing oil spills at sea. The degradation of alkylphenol polyethoxylates (non-ionic) can lead to the formation of alkylphenols (particularly nonylphenols), which act as endocrine disruptors.

High phosphate detergents such as tri-sodium phosphate (TSP) can be purchased at some paint and hardware stores. Regular cleaning with high phosphate detergents has proven to be effective in reducing lead dust. Lead dust accumulates in window wells and around doors or any other high friction surfaces.

Detergents can have poisonous effects in all types of aquatic life if they are present in sufficient quantities, and this includes the biodegradable detergents. All detergents destroy the external mucus layers that protect the fish from bacteria and parasites; plus they can cause severe damage to the gills. Most fish will die when detergent concentrations approach 15 parts per million. Detergent concentrations as low as 5 ppm will kill fish eggs. Surfactant detergents are implicated in decreasing the breeding ability of aquatic organisms.

Detergents also add another problem for aquatic life by lowering the surface tension of the water. Organic chemicals such as pesticides and phenols are then much more easily absorbed by the fish. A detergent concentration of only 2 ppm can cause fish to absorb double the amount of chemicals they would normally absorb, although that concentration itself is not high enough to affect fish directly.

Phosphates in detergents can lead to freshwater algal blooms that releases toxins and deplete oxygen in waterways. When the algae decompose, they use up the oxygen available for aquatic life.

The main contributors to the toxicity of detergents were the sodium silicate solution and the surfactants-with the remainder of the components contributing very little to detergent toxicity. The potential for acute aquatic toxic effects due to the release of secondary or tertiary sewage effluents containing the breakdown products of laundry detergents may frequently be low. However, untreated or primary treated effluents containing detergents may pose a problem. Chronic and/or other sublethal effects that were not examined in this study may also pose a problem.

4. Domestic Sewage And Farm Manures

5. Food Processing Wastes (Including Processes Taking Place On The Farm)

6. Gases (E.G. Chlorine, Ammonia)

Effluents are often complex mixtures of poisons. If two or more poisons are present together in an effluent they may exert a combined effect on an organism wich is additive. Some gases that can harm aquatic freshwater life are gases such as chlorine, ammonia and methane. Chlorine is very additive in combination with copper. It does not normally occur in the environment except as a yellow gas on rare occasions. It's a manufactured substance and the byproducts of chlorine (organochlorines and dioxins) are persistent in the environment. One of the largest uses of chlorine is in the paper industry. Chlorine is first used to break down the lignan that holds the wood fibers together. Then chlorine is used to bleach the paper to make it white.

The effluent or wastewater containing dioxins and other organochlorines are then dumped into streams and waterways. These ingredients are highly toxic and carcinogenic. Once in the waste stream, they come into contact with other organic materials and surfactants and combine to form a host of extremely toxic organic chemicals.

The water becomes polluted; the fish become contaminated; animals eat the fish and people eat the contaminated animals and fish.

It is so widespread that it would be difficult to find any human being who does not have detectable levels of dioxin in his/her blood.

Some environmentalists call for a ban on the use of chlorine as bleach in the pulp and the paper industry around the Great Lakes.

7. Heat

There are various effects on the biology of the ecosystems when heated effluents reach the receiving waters. The species that are intolerant to warm conditions may disappear, while others, rare in unheated water, may thrive so that the structure of the community changes. Thermal pollution can have a great influence on the aquatic ecosystem. Species that are restricted to heated waters, can build up large populations in the receiving waters. Respiration and growth rates may be changed and these may alter the feeding rates of organisms. The reproduction period may be brought forward and development may be speeded up. Parasites and diseases may also be affected.

An increase of temperature also means a decrease in oxygen solubility. Any reduction in the oxygen concentration of the water, particularly when organic pollution is also present, may result in the loss of sensitive species.

Possibly the most damaging environmental effect of a power station is the many organisms that may be sucked in through the water intake. Larger creatures, such as fish, are killed on the intake screens while smaller species pass through the plant. Even algae may be damaged, with permanent impairment of the photosynthetic mechanism. Liquid water changes temperature slowly because it can store a large amount of heat without a large change in temperature. This high heat capacity helps protect living organisms from temperature fluctuations, moderates the earth's climate and makes water an excellent coolant for car engines, power plants and heatproducing industrial processes. But when water is used in the industry, it is hot and it will be spilled through a discharge pipe into a river. This increase in temperature will reduce the amount of oxygen in the river. That can affect the level of oxygen freely available to organisms, which in turn affects respiration and essentially their way of life. For example, the metabolism rate is largely dependent upon the temperature of an animal's body. Animals display several different types of thermal adaptations to their environment. Two particularly prevalent types include ectotherms and endotherms. In ectotherms (an animal whose body temperature varies with the temperature of its surroundings; any animal except birds and mammals), the body temperature will be low in a cold environment and high in a warm environment. For example, in summer fish may have high metabolic rates because their body temperatures are elevated in the warm water. At the same time they are faced with relatively low oxygen availability because warm water holds less dissolved oxygen than cold water. The interaction of these factors may prove critical. For this reason there is a growing concern among ecologists about the heating of aquatic habitats by effluents from industrial and nuclear generating facilities. Heated water can kill animals and plants that are accustomed to living at lower temperatures.

8. Metals (e.g. cadmium, lead, mercury)

Three countries—the United States, Germany, and Russia—with only 8% of the world's population consume about 75% of the world's most widely used metals. The United States, with 4.5% of the world's population, uses about 20% of the worlds metal population and 25% of the fossil fuels produced each year.

Metals are introduced in aquatic systems as a result of the weathering of soils and rocks, from volcanic eruptions, and from a variety of human activities involving the mining, processing, or use of metals and/or substances that contain metal pollutants. The most common heavy metal pollutants are arsenic, cadmium, chromium, copper, nickel, lead and mercury. There are different types of sources of pollutants: point sources (localized pollution), where pollutants come from single, identifiable sources. The second type of pollutant sources are nonpoint sources, where pollutants come from dispersed (and often difficult to identify) sources. There are only a few examples of localized metal pollution, like the natural weathering of ore bodies and the little metal particles coming from coal-burning power plants via smokestacks in air, water and soils around the factory.

The most common metal pollution in freshwater comes from mining companies. They usually use an acid mine drainage system to release heavy metals from ores, because metals are very soluble in an acid solution. After the drainage process, they disperse the acid solution in the groundwater, containing high levels of metals.

The term heavy metal is somewhat imprecise, but includes most metals with an atomic number greater than 20, and excludes alkali metals, alkaline earths, lanthanides and actinides.

When the pH in water falls, metal solubility increases and the metal particles become more mobile. That is why metals are more toxic in soft waters. Metals can become locked up in bottom sediments, where they remain for many years. Streams coming from draining mining areas are often very acidic and contain high concentrations of dissolved metals with little aquatic life. Both localized and dispersed metal pollution cause environmental damage because metals are non-biodegradable. Unlike some organic pesticides, metals cannot be broken down into less harmful components in the environment.

Campbell and Stokes (1985) described two contrasting responses of an organism to a metal toxicity with declining pH:

- If there is little change in speciation and the metal binding is weak at the biological surface, a decrease in pH will decrease owning to competition for binding sites from hydrogen ions.

- Where there is a marked effect on speciation and strong binding of the metal at the biological surface, the dominant effect of a decrease in pH will be to increase the metal availability.

Generally the ionic form of a metal is more toxic, because it can form toxic compounds with other ions. Electron transfer reactions that are connected with oxygen can lead to the production of toxic oxyradicals, a toxicity mechanism now known to be of considerable importance in both animals and plants. Some oxyradicals, such as superoxide anion (O2-) and the hydroxyl radical (OH-), can cause serious cellular damage.

Some inorganic pollutants are assimilated by organisms to a greater extent than others. This is reflected in the Bioconcentration Factor (BCF), which can be expressed as follows:

BCF = concentration of the chemical in the organism / concentration of the chemical in the ambient environment.

The ambient environment for aquatic organisms is usually the water or sediments. With inorganic chemicals, the extent of long-term bioaccumulation depends on the rate of excretion. Toxic chemicals can be stored into tissues of species, especially fat tissues. Bioaccumulation of cadmium in animals is high compared to most of the other metals, as it is assimilated rapidly and excreted slowly. Also the sensitivity of individuals of a particular species to a pollutant may be influenced by factors such as sex, age, or size. In general the concentrations of metals in invertebrates are inversely related to their body mass. In fish, the embryonic and larval stages are usually the most sensitive to pollutants.

Benthic organisms are likely to be the most directly affected by metal concentrations in the sediments, because the benthos is the ultimate repository of the particulate materials that are washed into aquatic systems.

Metal Tolerance

Some metals, such as manganese, iron, copper, and zinc are essential micronutrients. They are essential to life in the right concentrations, but in excess, these chemicals can be poisonous. At the same time, chronic low exposures to heavy metals can have serious health effects in the long run.

Tolerance to metals has also been recorded in invertebrates and in fish. After exposure for 24 hours to a copper concentration of 0.55 mg/l, rainbow trout showed a 55 per cent inhibition of sodium uptake and a 4 per cent reduction in affinity for sodium, which resulted in an overall decrease in total sodium concentration of sulphydryl-rich protein (Lauren and McDonald 1987a,b). The protein was considered to be a metallothionein. These low molecular weight proteins contain many sulphur-rich amino acids which bind and detoxify some metals. The pretreatment of an organism with low doses of a metal may stimulate metallothionein synthesis and provide tolerance during a subsequent exposure (Pascoe and Beattie, 1979).

Many rivers are polluted with heavy metals from old mine workings and some species of algae become very tolerant to polluted conditions. A survey of 47 sites with different concentration of zinc found the filamentous green alga 'Hormidium rivulare' to be abundant everywhere, tolerating zinc concentrations as high as 30.2 mg Zn/l.

Toxicity of Metals

For the protection of human health, the maximum permissible concentrations for metals in natural waters that are recommended by the Environmental Protection Agency (EPA), are listed below:

Maximum Permissible Concentrations (MPC) of Various Metals in Natural Waters For the Protection of Human Health

Metal	Chemical Symbol	Mg m⁻³
Mercury	Hg	0.144
Lead	Pb	5
Cadmium	Cd	10
Selenium	Se	10
Thallium	Tl	13
Nickel	Ni	13.4
Silver	Ag	50
Manganese	Mn	50
Chromium	Cr	50
Iron	Fe	300
Barium	Ba	1000

This table gives an idea of the relative toxicity of various metals. Mercury, lead and cadmium are not required even in small amounts by any organism.

Because metals are rather insoluble in neutral or basic pH, pHs of 7 or above give a highly misleading picture of the degree of metal pollution. So in some cases it may underestimate significantly the total of metal concentrations in natural waters.

9. Nutrients (especially phosphates, nitrates)

Aquatic plants (like any other plants) need two essential nutrients to grow: nitrogen (N) and phosphorus (P). In a healthy lake the nutrients occur in small amounts. But in large quantities, they can cause a major water pollution problem. Too many nutrients stimulate the rapid growth of plants and algae, clogging waterways and sometimes creating blooms of toxic blue-green algae. This process is called eutrophication. The result of this is that when the plants and algae die and decompose, they use up large amounts of oxygen (O2). So the amount of oxygen that is available for fish and other aquatic species will be reduced. In extreme cases it can lead to a completely oxygenl ess environment that can support nothing except a few species of anaerobic bacteria. It also can kill fish and other aquatic life and reduce the aesthetic and recreational value of the lake.

The nutrients include nitrates found in sewage and fertilizers, and phosphates found in detergents and fertilizers. Human inputs of nutrients from the atmosphere and from nearby urban and agricultural areas can accelerate the natural eutrophication of lakes, a process called cultural eutrophication.

Nutrients from urban sources may be derived from domestic sewage, industrial wastes and storm drainage. The contribution of nitrogen and phosphorus per person per day averages 10.8 g N and 2.2 g P, though there is a considerable range. In the 1940s detergents were developed containing sodium tripolyphosphate, which softens water by neutralizing calcium and keeps dirt in suspension once it has washed off clothes. These are the principal sources of nutrient overload causing cultural eutrophication in lakes. The amount of each source varies according to the types and amounts of human activities occurring in each airshed and watershed.

10. Oil And Oil Dispersants

11. Organic Toxic Wastes (e.g. formaldehyde, phenols)

Organic pollution occurs when large quantities of organic compounds, which act as substrates for microorganisms, are released into watercources. During the decomposition process the dissolved oxygen in the receiving water may be used up at a greater rate than it can be replenished, causing oxygen depletion and having severe consequences for the stream biota. Organic effluents also frequently contain large quantities of suspendid solids which reduce the light available to photosynthetic organisms and, on settling out, alter the characteristics of the river bed, rendering it an unsuitable habitat for many invertebrates. Toxic ammonia is often present.

Organic pollutants consist of proteins, carbohydrates, fats and nucleic acids in a multiplicity of combinations. Raw sewage is 99,9 per cent water, and of the 0,1 per cent solids, 70 per cent is organic (65 per cent proteins, 25 per cent carbohydrates, 10 per cent fats). Organic wastes from people and their animals may also be rich in disease-causing (pathogenic) organisms.

Origins of Organic Pollutants

Organic pollutants originate from domestic sewage (raw or treated), urban run-off, industrial (trade) effluents and farm wastes. Sewage effluents is the greatest source of organic materials discharged to freshwaters. In England and Wales there are almost 9000 discharges releasing treated sewage effluent to rivers and canals and several hundred more discharges of crude sewage, the great majority of them tot the lower, tidal reaches of rivers or, via long outfalls, to the open sea. It has been assumed, certainly incorrectly, that the sea has an almost unlimited capacity for purifying biodegradable matter.

The effects of Organic Effluents on Receiving Waters

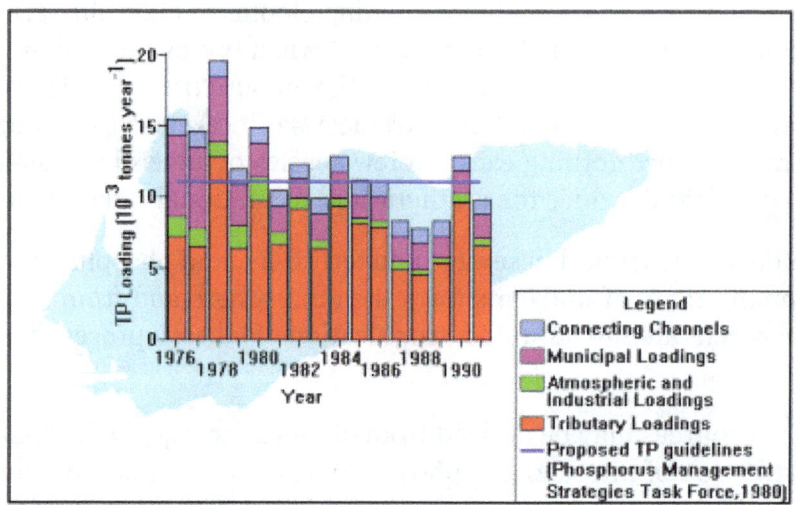

When an organic polluting load id discharged into a river it is gradually eliminated by the activities of micro organisms in a way very similar to the processes in the sewage treatment works. This self-purification requires sufficient concentrations of oxygen, and involves the breakdown of complex organic molecules into simple in organic molecules. Dilution, sedimentation and sunlight also play a part in the process. attached micro organisms in streams play a greater role than suspended organisms in self-purification. Their importance increases as the quality of the effluent increases since attached microorganisms are already present in the stream, whereas suspended ones are mainly supplied with the discharge.

Effects on the Biota

Organic pollution affects the organisms living in a stream by lowering the available oxygen in the water. This causes reduced fitness, or, when severe, asphyxiation. The increased turbidity of the water reduces the light available to photosynthetic organisms. Organic wastes also settle out on the bottom of the stream, altering the characteristics of the substratum.

12. Pathogens

A pathogen is an organism that produces a disease. It is an organic pollution (biological hazard) and occurs from fecal contaminations. Fecal contaminations of water can introduce a variety of pathogens into waterways, including bacteria, viruses, protozoa and parasitic worms.

Bacteria

A very well known pathogenic bacteria is Salmonella. There are some 200 immunologically distinguishable types of Salmonella known to be pathogenic to humans. But there are many more that infect animals, including livestock. Crossinfection between people can occur via water pollution. The spreading of untreated sewage wastewater on land and its use for the irrigation of crops can also be a source of infection.

Currently there appears to be an increase in the spread of Salmonella and this has been related to modern living conditions, such as mass food production (e.g. of poultry) and communal feeding.

13. Pesticides

14. Polychlorinated Biphenyls

15. Radionuclides

Assessment Of Freshwater Pollution:

The —Conventional Pollutant Measures are:

- Oxygen (BOD, COD, DO)

- Solids content (TSS, Conductivity, Secchi disk, settleable solids)

- Nutrients (phosphorus, nitrogen) /Algae/Eutrophication

- Acidity (pH)

- Bacteria (e.g., fecal coliform)

- Temperature

Major approaches to Biomonitoring/assessment in the last 30 years:

- single metric – biotic indices

- multivariate (predictive models (RIVPACS, AUSRIVAS, BEAST))

- multimetric indices-index of biological integrity IBI & AQEM system

Streams can become polluted by water entering from agricultural land or industrial sites, and the quality of the water will be reflected by the types of creatures that can survive. Some of the worms can manage in very heavily polluted water (Tubifex is a good example), while some of the insects require clean conditions.

Biological Indicators

Bilogical indicators can be defined as organism/organisms or attributes of the community which can be used to provide information on:

- state of the environment

- change from 'normal conditions'

- Highlight the pressure causing a change.

Characteristics of an Ideal Indicator

- taxonomically sound and easy to identify

- widespread distribution

- numerically abundant

- large body size

- ecological requirements known (autecology)

- narrow ecological demands

Types of Indicators

- sentinel organisms

- community level indicators

- organism level indicators

  - biochemical indicators

  - life history response

  - morphological deformities egs.

In order to stimulate industrial development and economic growth, many African countries have relaxed their anti-pollution regulation where it existed and environmental laws are rarely a constraint. The consequence is industrial pollution problems. The detection of agro industrial pollution is urgent in Africa to insure the protection of water resources against organic, inorganic pollutants and pesticides which are released in freshwater ecosystems. Methods for the detection of water pollution are chemical and biological. Chemical analyses do not detect punctual pollutions, are costly and not sustainable for third world countries. Several organisms are used as water quality indicators. These include bacteria, ciliates, diatoms and macroinvertebrates. Macroinvertebrates are potential markers of water pollution as some members are sensitive to different levels of pollution. The abundance of these organisms, their wide distribution, the ease with which they are identified, and their sedentarity have contributed to their utilization in water pollution assessment. Most reports of the use of macroinvertebrates in freshwater pollution monitoring are from Europe, North America and Asia. These data are rare in Africa which is biogeographically different. Studies of freshwater organisms for public health purposes have mainly focused on medically important snails. The development of sampling technique which can bring both molluscs and others macroinvertebrates is necessary in the biological evaluation of freshwater pollution.

Macroinvertebrates

Macroinvertebrates are a key indicator group. The larger freshwater invertebrates (macroinvertebrates) spend most of their time in the same part of the stream. They are numerous and easy to catch, and once caught and identified they can give clues to the pollution history of that particular part of the stream. By sampling different locations it is possible to locate the source of any polluting incident – upstream will not be damaged, but downstream will have been affected.

Biotic Index

There is a general rule that better water has a larger number of different invertebrate species, and by counting the number of different types of creatures it is possible to get a rough idea of water quality. By going further and identifying each species found it is possible to quantify this information. It is possible to arrive at the Biotic Index.

Advantages of macroinvertebrates in water quality assessment

- wide diversity (species & functional groups) and abundance

- relatively sedentary-occurrence of most can be related to conditions at place of capture

- life cycle of 6 months or longer-provides overview of prevailing physical/chemical conditions

- sampling is relatively easy and cheap

- they respond to environmental stress-integrate the effects of short-term perturbations.

Disadvantages of using macroinvertebrates in water quality assessment

- biological expertise is needed to identify some groups

- autoecology of various groups needed as absences may be related to habitat or life cycle factors

Macroinvertebrates respond to environmental stress conditions such as:

- oxygen depletion

- direct toxicity

- loss of maicrohabitat

- siltation of habitat

- food availability changes

- competition from other species

Process of bioassessment using macroinvertebrates involves the following steps:

- sample collection

- sorting

- identification

- data analysis

- reporting

Chemical Analysis

The alternative way of studying water pollution is by direct chemical analysis, and while it will be specific about the nature of the pollutants it can only help when the stream actually contains them. Once the damage has swept downstream the water will be replaced, but the invertebrates living in the affected area will have been harmed, and they will take a long time to recover. This means that occasional sampling of stream macroinvertebrates can be used to discover pollution events that occurred in the past.

Indicator Species

If we are able to recognise one or two of the 'indicator species' you will not have to go to the trouble of identifying all the other creatures. If Stonefly larvae are present then the water is good, if they are absent but there are still Mayfly larvae then the water is reasonable. If neither of these is present then there has probably been a problem and a more detailed study might be called for.

Environmental Impact of Fishing

The environmental impact of fishing includes issues such as the availability of fish, overfishing, fisheries, and fisheries management; as well as the impact of fishing on other elements of the environment, such as by-catch. These issues are part of marine conservation, and are addressed in fisheries science programs. There is a growing gap between the supply of fish and demand, due in part to world population growth. Similar to other environmental issues, there can be conflict between the fishermen who depend on fishing for their income, and fishery scientists whose studies indicate that if future fish populations are to be sustainable then some fisheries must reduce or even close.

Fishing down the foodweb

The journal *Science* published a four-year study in November 2006, which predicted that, at prevailing trends, the world would run out of wild-caught seafood in 2048. The scientists stated that the decline was a result of overfishing, pollution and other environmental factors that were reducing the population of fisheries at the same time as their ecosystems were being annihilated. Yet again the analysis has met criticism as being fundamentally flawed, and many fishery management officials, industry representatives and scientists challenge the findings, although the debate continues. Many countries, such as Tonga, the United States, Australia and Bahamas, and international management bodies have taken steps to appropriately manage marine resources.

Effects on Marine Habitat

A sea turtle killed by a boat propeller

Some fishing techniques also may cause habitat destruction. Blast fishing and cyanide fishing, which are illegal in many places, harm surrounding habitat. Bottom trawling, the practice of pulling a fishing net along the sea bottom behind trawlers, removes around 5 to 25% of an area's seabed life on a single run. A 2005 report of the UN Millennium Project, commissioned by UN Secretary-General Kofi Annan, recommended the elimination of bottom trawling on the high seas by 2006 to protect seamounts and other ecologically sensitive habitats. This was not done.

In mid-October 2006, US President Bush joined other world leaders calling for a moratorium on deep-sea trawling, a practice shown to often have harmful effects on sea habitat and, hence, on fish populations. No further action was taken (Divek).

Overfishing

Overfishing has also been widely reported due to increases in the volume of fishing hauls to feed a quickly growing number of consumers. This has led to the breakdown of some sea ecosystems and several fishing industries whose catch has been greatly diminished. The extinction of many species has also been reported. According to a Food and Agriculture Organization estimate, over 70% of the world's fish species are either fully exploited or depleted. According to the Secretary General of the 2002 World Summit on Sustainable Development, "Overfishing cannot continue, the depletion of fisheries poses a major threat to the food supply of millions of people."

The cover story of the May 15, 2003 issue of the science journal *Nature* – with Dr. Ransom A. Myers, an internationally prominent fisheries biologist (Dalhousie University, Halifax, Canada) as the lead author – was devoted to a summary of the scientific information. The story asserted that, as compared with 1950 levels, only a remnant (in some instances, as little as 10%) of all large ocean-fish stocks are left in the seas. These large ocean fish are the species at the top of the food chains (e.g., tuna, cod, among others). This article was subsequently criticized as being fundamentally flawed, although much debate still exists (Walters 2003; Hampton et al. 2005; Maunder et al. 2006; Polacheck 2006;Sibert et al. 2006) and the majority of fisheries scientists now consider the results irrelevant with respect to large pelagics (the open seas).

Ecological Disruption

Fishing may disrupt food webs by targeting specific, in-demand species. There might be too much fishing of prey species such as sardines and anchovies, thus reducing the food supply for the predators. It may also cause the increase of prey species when the target fishes are predator species such as salmon and tuna.

Bycatch

Bycatch is the portion of the catch that is not the target species. These are either kept to be sold or discarded. In some instances the discarded portion is known as discards. Even sports fisherman discard a lot of non-target and target fish on the bank while fishing.

Marine Debris

Recent research has shown that fishing debris such as nets, buoys, and lines, accounts for a majority of plastic debris found in the oceans, such as in the Great Pacific garbage patch. Similarly, fishing debris has been shown to be a major source of plastic debris found on the shores of Korea.

Possible Remedies

Many governments and intergovernmental bodies have implemented fisheries management policies designed to curb the environmental impact of fishing. Fishing conservation aims to control the human activities that may completely decrease a fish stock or washout an entire aquatic environment. These laws include the quotas on the total catch of particular species in a fishery, effort quotas (e.g., number of days at sea), the limits on the number of vessels allowed in specific areas, and the imposition of seasonal restrictions on fishing.

In 2008 a large scale study of fisheries that used individual transferable quotas and ones that didn't provided strong evidence that individual transferable quotas can help to prevent collapses and restore fisheries that appear to be in decline.

Fish farming has been proposed as a more sustainable alternative to traditional capture of wild fish. However, fish farming has been found to have negative impacts on nearby wild fish and farming of predatory fish like salmon can rely on fish feed that is based on fish meal and oil from wild fish.

The environmental impact of recreational fishing may be alleviated to some extent by catch and release fishing.

Environmental Impact of Shipping

The environmental impact of shipping includes greenhouse gas emissions, acoustic, and oil pollution. The International Maritime Organization (IMO) estimates that Carbon dioxide emissions from shipping were equal to 2.2% of the global human-made emissions in 2012 and expects them to rise by as much as 2 to 3 times by 2050 if no action is taken.

The First Intersessional Meeting of the IMO Working Group on Greenhouse Gas Emissions from Ships took place in Oslo, Norway on 23–27 June 2008. It was tasked with developing the technical basis for the reduction mechanisms that may form part of a future IMO regime to control greenhouse gas emissions from international shipping, and a draft of the actual reduction mechanisms themselves, for further consideration by IMO's Marine Environment Protection Committee (MEPC).

A cargo ship discharging ballast water into the sea.

Ballast Water

Ballast water discharges by ships can have a negative impact on the marine environment.

Cruise ships, large tankers, and bulk cargo carriers use a huge amount of ballast water, which is often taken on in the coastal waters in one region after ships discharge wastewater or unload cargo, and discharged at the next port of call, wherever more cargo is loaded. Ballast water discharge typically contains a variety of biological materials, including plants, animals, viruses, and bacteria. These materials often include non-native, nuisance, invasive, exotic species that can cause extensive ecological and economic damage to aquatic ecosystems along with serious human health problems.

Sound Pollution

Noise pollution caused by shipping and other human enterprises has increased in recent history. The noise produced by ships can travel long distances, and marine species who may rely on sound for their orientation, communication, and feeding, can be harmed by this sound pollution

The Convention on the Conservation of Migratory Species has identified ocean noise as a potential threat to marine life. The disruption of whales' ability to communicate with one another is an extreme threat and is affecting their ability to survive. According to Discovery Channel's article on Sonic Sea Journeys Deep Into the Ocean, over the last century, extremely loud noise from commercial ships, oil and gas exploration, naval sonar exercises and other sources has transformed the ocean's delicate acoustic habitat, challenging the ability of whales and other marine life to prosper and ultimately to survive. Whales are starting to react to this in ways that are life-threatening. Kenneth C. Balcomb, a whale researcher and a former U.S Navy officer, states that the day March 15, 2000, is the day of infamy. As Discovery says, where him and his crew discovered whales swimming dangerously close to the shore. They're supposed to be in deep water. So I pushed it back out to sea, says Balcomb. Although sonar helps to protect us, it is destroying marine life. According to IFAW Animal Rescue Program Director Katie Moore, "There's different ways that sounds can affect animals. There's that underlying ambient noise level that's rising, and rising, and rising that interferes with communication and their movement patterns. And then there's the more acute kind of traumatic impact of sound, that's causing physical damage or a really strong behavioral response. It's fight or flight".

Wildlife Collisions

Marine mammals, such as whales and manatees, risk being struck by ships, causing injury and death. For example, if a ship is traveling at a speed of only 15 knots, there is a 79 percent chance of a collision being lethal to a whale.

One notable example of the impact of ship collisions is the endangered North Atlantic right whale, of which 400 or less remain. The greatest danger to the North Atlantic right whale is injury sustained from ship strikes. Between 1970 and 1999, 35.5 percent of recorded deaths were attributed to collisions. During 1999 to 2003, incidents of mortality and serious injury attributed to ship strikes averaged one per year. In 2004 to 2006, that number increased to 2.6. Deaths from collisions has become an extinction threat.

Atmospheric Pollution

Exhaust gases from ships are considered to be a significant source of air pollution, both for conventional pollutants and greenhouse gases.

There is a perception that cargo transport by ship is low in air pollutants, because for equal weight and distance it is the most efficient transport method, according to shipping researcher Amy Bows-Larkin. This is particularly true in comparison to air freight; however, because sea shipment accounts for far more annual tonnage and the distances are often large, shipping's emissions are globally substantial. A difficulty is that the year-on-year increasing amount shipping overwhelms gains in efficiency, such as from slow-steaming or the use of kites. The growth in tonne-kilometers of sea shipment has averaged 4 percent yearly since the 1990s. And it has grown by a factor of 5 since the 1970s. There are now over 100,000 transport ships at sea, of which about 6,000 are large container ships.

Conventional Pollutants

Air pollution from cruise ships is generated by diesel engines that burn high sulfur content fuel oil, also known as bunker oil, producing sulfur dioxide, nitrogen oxide and particulate, in addition to carbon monoxide, carbon dioxide, and hydrocarbons. Diesel exhaust has been classified by EPA as a likely human carcinogen. EPA recognizes that these emissions from marine diesel engines contribute to ozone and carbon monoxide non-attainment (i.e., failure to meet air quality standards), as well as adverse health effects associated with ambient concentrations of particulate matter and visibility, haze, acid deposition, and eutrophication and nitrification of water. EPA estimates that large marine diesel engines accounted for about 1.6 percent of mobile source nitrogen oxide emissions and 2.8 percent of mobile source particulate emissions in the United States in 2000. Contributions of marine diesel engines can be higher on a port-specific basis. Ultra-low sulfur diesel (ULSD) is a standard for defining diesel fuel with substantially lowered sulfur contents. As of 2006, almost all of the petroleum-based diesel fuel available in Europe and North America is of a ULSD type.

In 2016 the IMO has made new sulfur regulations which must be implemented by larger ships by 2020.

Of total global air emissions, shipping accounts for 18 to 30 percent of the nitrogen oxide and 9 percent of the sulphur oxides. Sulfur in the air creates acid rain which damages crops and buildings. When inhaled, sulfur is known to cause respiratory problems and even increases the risk of a heart attack. According to Irene Blooming, a spokeswoman for the European environmental coalition Seas at Risk, the fuel used in oil tankers and container ships is high in sulfur and cheaper to buy compared to the fuel used for domestic land use. "A ship lets out around 50 times more sulfur than a lorry per metric tonne of cargo carried." Cities in the U.S. like Long Beach, Los Angeles, Houston, Galveston, and Pittsburgh see some of the heaviest shipping traffic in the nation and have left local officials desperately trying to clean up the air. Increasing trade between the U.S. and China is helping to increase the number of vessels navigating the Pacific and exacerbating many of the environmental problems. To maintain the level of growth China is experiencing, large amounts of grain are being shipped to China by the boat load. The number of voyages are expected to continue increasing.

Greenhouse Gas Pollutants

Cruise ship haze over Juneau, Alaska

3.5 to 4 percent of all climate change emissions are caused by shipping, primarily carbon dioxide.

As one way to reduce the impact of greenhouse gas emissions from shipping, vetting agency Right-Ship developed an online "Greenhouse Gas (GHG) Emissions Rating" as a systematic way for the industry to compare a ship's CO_2 emissions with peer vessels of a similar size and type. Based on the International Maritime Organisation's (IMO) Energy Efficiency Design Index (EEDI) that applies to ships built from 2013, RightShip's GHG Rating can also be applied to vessels built prior to 2013, allowing for effective vessel comparison across the world's fleet. The GHG Rating utilises an A to G scale, where A represents the most efficient ships. It measures the theoretical amount of carbon dioxide emitted per tonne nautical mile travelled, based on the design characteristics of the ship at time of build such as cargo carrying capacity, engine power and fuel consumption. Higher rated ships can deliver significantly lower CO_2 emissions across the voyage length, which means they also use less fuel and are cheaper to run.

Stress for Improvement

One source of environmental stresses on maritime vessels recently has come from states and localities, as they assess the contribution of commercial marine vessels to regional air quality problems when ships are docked at port. For instance, large marine diesel engines are believed to contribute 7 percent of mobile source nitrogen oxide emissions in Baton Rouge/New Orleans. Ships can also have a significant impact in areas without large commercial ports: they contribute about 37 percent of total area nitrogen oxide emissions in the Santa Barbara area, and that percentage is expected to increase to 61 percent by 2015. Again, there is little cruise-industry specific data on this issue. They comprise only a small fraction of the world shipping fleet, but cruise ship emissions may exert significant impacts on a local scale in specific coastal areas that are visited repeatedly. Shipboard incinerators also burn large volumes of garbage, plastics, and other waste, producing ash that must be disposed off. Incinerators may release toxic emissions as well.

In 2005, MARPOL Annex VI came into force to combat this problem. As such cruise ships now employ CCTV monitoring on the smokestacks as well as recorded measuring via opacity meter while some are also using clean burning gas turbines for electrical loads and propulsion in sensitive areas.

Oil Spills

Most commonly associated with ship pollution are oil spills. While less frequent than the pollution that occurs from daily operations, oil spills have devastating effects. While being toxic to marine life, polycyclic aromatic hydrocarbons (PAHs), the components in crude oil, are very difficult to clean up, and last for years in the sediment and marine environment. Marine species constantly exposed to PAHs can exhibit developmental problems, susceptibility to disease, and abnormal reproductive cycles. One of the more widely known spills was the Exxon Valdez incident in Alaska. The ship ran aground and dumped a massive amount of oil into the ocean in March 1989. Despite efforts of scientists, managers and volunteers, over 400,000 seabirds, about 1,000 sea otters, and immense numbers of fish were killed.

International Regulation

Some of the major international efforts in the form of treaties are the Marine Pollution Treaty, Honolulu, which deals with regulating marine pollution from ships, and the UN Convention on Law of the Sea, which deals with marine species and pollution. While plenty of local and international regulations have been introduced throughout maritime history, much of the current regulations are considered inadequate. "In general, the treaties tend to emphasize the technical features of safety and pollution control measures without going to the root causes of sub-standard shipping, the absence of incentives for compliance and the lack of enforceability of measures." The most common problems encountered with international shipping arise from paperwork errors and customs brokers not having the proper information about your items. Cruise ships, for example, are exempt from regulation under the US discharge permit system (NPDES, under the Clean Water Act) that requires compliance with technology-based standards. In the Caribbean, many ports lack proper waste disposal facilities, and many ships dump their waste at sea. Moreover, due to the complexities of shipping trade and the difficulties involved in regulating this business, a comprehensive and generally acceptable regulatory framework on corporate responsibility for reducing GHG emissions is unlikely to be achieved soon. In fact, emissions are continuing to increase. Under these circumstances, it is necessary for the states, the shipping industry and global organizations to explore and discuss market based mechanisms for vessel-sourced GHG emissions reduction.

Sewage

Carcass of a whale on a shore in Iceland.

The cruise line industry dumps 255,000 US gallons (970 m³) of greywater and 30,000 US gallons (110 m³) of blackwater into the sea every day. Blackwater is sewage, wastewater from toilets and medical facilities, which can contain harmful bacteria, pathogens, viruses, intestinal parasites, and harmful nutrients. Discharges of untreated or inadequately treated sewage can cause bacterial and viral contamination of fisheries and shellfish beds, producing risks to public health. Nutrients in sewage, such as nitrogen and phosphorus, promote excessive algal blooms, which consumes oxygen in the water and can lead to fish kills and destruction of other aquatic life. A large cruise ship (3,000 passengers and crew) generates an estimated 55,000 to 110,000 liters per day of blackwater waste.

Due to the environmental impact of shipping, and sewage in particular marpol annex IV was brought into force September 2003 strictly limiting untreated waste discharge. Modern cruise ships are most commonly installed with a membrane bioreactor type treatment plant for all blackwater and greywater, such as , Zenon or Rochem which produce near drinkable quality effluent to be re-used in the machinery spaces as technical water.

Cleaning

Greywater is wastewater from the sinks, showers, galleys, laundry, and cleaning activities aboard a ship. It can contain a variety of pollutant substances, including fecal coliforms, detergents, oil and grease, metals, organic compounds, petroleum hydrocarbons, nutrients, food waste, medical and dental waste. Sampling done by the EPA and the state of Alaska found that untreated greywater from cruise ships can contain pollutants at variable strengths and that it can contain levels of fecal coliform bacteria several times greater than is typically found in untreated domestic wastewater. Greywater has potential to cause adverse environmental effects because of concentrations of nutrients and other oxygen-demanding materials, in particular. Greywater is typically the largest source of liquid waste generated by cruise ships (90 to 95 percent of the total). Estimates of greywater range from 110 to 320 liters per day per person, or 330,000 to 960,000 liters per day for a 3,000-person cruise ship.

Solid Waste

Solid waste generated on a ship includes glass, paper, cardboard, aluminium and steel cans, and plastics. It can be either non-hazardous or hazardous in nature. Solid waste that enters the ocean may become marine debris, and can then pose a threat to marine organisms, humans, coastal communities, and industries that utilize marine waters. Cruise ships typically manage solid waste by a combination of source reduction, waste minimization, and recycling. However, as much as 75 percent of solid waste is incinerated on board, and the ash typically is discharged at sea, although some is landed ashore for disposal or recycling. Marine mammals, fish, sea turtles, and birds can be injured or killed from entanglement with plastics and other solid waste that may be released or disposed off of cruise ships. On average, each cruise ship passenger generates at least two pounds of non-hazardous solid waste per day. With large cruise ships carrying several thousand passengers, the amount of waste generated in a day can be massive. For a large cruise ship, about 8 tons of solid waste are generated during a one-week cruise. It has been estimated that 24 percent of the solid waste generated by vessels worldwide (by weight) comes from cruise ships. Most cruise ship garbage is treated on board (incinerated, pulped, or ground up) for discharge overboard. When garbage must be off-loaded (for example, because glass and aluminium cannot be incinerated), cruise ships can put a strain on port reception facilities, which are rarely adequate to the task of serving a large passenger vessel.

Bilge Water

On a ship, oil often leaks from engine and machinery spaces or from engine maintenance activities and mixes with water in the bilge, the lowest part of the hull of the ship, but there is a filter to clean bilge water before being discharged. Oil, gasoline, and by-products from the biological breakdown of petroleum products can harm fish and wildlife and pose threats to human health if ingested. Oil in even minute concentrations can kill fish or have various sub-lethal chronic effects. Bilge water also may contain solid wastes and pollutants containing high levels of oxygen-demanding material, oil and other chemicals. A typically large cruise ship will generate an average of 8 metric tons of oily bilge water for each 24 hours of operation. To maintain ship stability and eliminate potentially hazardous conditions from oil vapors in these areas, the bilge spaces need to be flushed and periodically pumped dry. However, before a bilge can be cleared out and the water discharged, the oil that has been accumulated needs to be extracted from the bilge water, after which the extracted oil can be reused, incinerated, and/or offloaded in port. If a separator, which is normally used to extract the oil, is faulty or is deliberately bypassed, untreated oily bilge water could be discharged directly into the ocean, where it can damage marine life. A number of cruise lines have been charged with environmental violations related to this issue in recent years.

Issues by Region

European Union

- EU Reducing Greenhouse Gas emissions from the shipping sector

- EU Sustainable Shipping Forum (ESSF)

- EC-IMO Energy Efficiency Project. The 4-year project aims to establish Maritime Technology Cooperation Centres in 5 regions: Africa, Asia, the Caribbean, Latin America and the Pacific. Through technical assistance and capacity-building, the centres will promote the uptake of low carbon technologies and operations in maritime transport in the less developed countries in the respective region. This will also support the implementation of the internationally agreed energy efficiency rules and standards (EEDI and SEEMP).

- EEDI=Energy Efficiency Design Index

- SEEMP=Ship Energy Efficiency Management Plan

- MRV Monitoring, reporting and verification of CO_2 emissions from large ships using EU ports

United Kingdom

- Merchant Shipping Act 1995

- Merchant Shipping (Pollution) Act 2006

United States

It is expected that, (from 2004) "...shipping traffic to and from the United States is projected to double by 2020."

- Act to Prevent Pollution from Ships

- American Bureau of Shipping

- Cruise ship pollution in the United States

- National Oil and Hazardous Substances Contingency Plan

- Oil Pollution Act of 1990

- Regulation of ship pollution in the United States

International Regulation

- MARPOL 73/78

Oil Pollution Toxicity to Marine Fish

Oil pollution toxicity to marine fish has been observed from oil spills such as the *Exxon Valdez* disaster, and from nonpoint sources, such as surface runoff, which is the largest source of oil pollution in marine waters. Crude oil entering waterways from spills or runoff contain polycyclic aromatic hydrocarbons (PAHs), the most toxic components of oil. The route of PAH uptake into fish depends on many environmental factors and the properties of the PAH. The common routes are ingestion, ventilation of the gills, and dermal uptake. Fish exposed to these PAHs exhibit an array of toxic effects including genetic damage, morphological deformities, altered growth and development, decreased body size, inhibited swimming abilities and mortality. The morphological deformities of PAH exposure, such as fin and jaw malformations, result in significantly reduced survival in fish due to the reduction of swimming and feeding abilities. While the exact mechanism of PAH toxicity is unknown, there are four proposed mechanisms. The difficulty in finding a specific toxic mechanism is largely due to the wide variety of PAH compounds with differing properties.

History

Research on the environmental impact of the petroleum industry began in earnest, during the mid to late 20th century, as the oil industry developed and expanded. Large scale transport of crude oil increased as a result of the increasing worldwide demand for oil, subsequently increasing the number of oil spills. Oil spills provided perfect opportunities for scientists to examine the in situ effects of crude oil exposure to marine ecosystems, and collaborative efforts between the National Oceanic and Atmospheric Administration (NOAA) and the United States Coast Guard resulted in improved response efforts and detailed research on oil pollution's effects. The Exxon Valdez oil spill in 1989, and the Deepwater Horizon oil spill in 2010, both resulted in increased scientific knowledge on the specific effects of oil pollution toxicity to marine fish.

Exxon Valdez Oil Spill

Focused research on oil pollution toxicity to fish began in earnest in 1989, after the *Exxon Valdez* tanker struck a reef in Prince William Sound, Alaska and spilled approximately 11 million gallons of crude oil into the surrounding water. At the time, the Exxon Valdez oil spill was the largest in the history of the United States. There were many adverse ecological impacts of the spill including the loss of the loss of billions of Pacific herring and pink salmon eggs. Pacific herring were just beginning to spawn in late March when the spill occurred, resulting in nearly half of the populations eggs being exposed to crude oil. Pacific herring spawn in the intertidal and subtidal zones, making the vulnerable eggs easily exposed to pollution.

Deepwater Horizon Oil Spill

After April 20, 2010, when an explosion on the *Deepwater Horizon* Macondo oil drilling platform triggered the largest oil spill in US history, another opportunity for oil toxicity research was presented. Approximately 171 million gallons of crude oil flowed from the seafloor into the Gulf of Mexico, exposing the majority of surrounding biota. The Deepwater Horizon oil spill also coincided directly with spawning window of various ecologically and commercially important fish species, including yellowfin and Atlantic bluefin tuna. The oil spill directly affected Atlantic bluefin tuna, as approximately 12% of larval tuna were located in oil contaminated waters, and Gulf of Mexico is the only known spawning grounds for the western population of bluefin tuna.

Exposure to Oil

Oil spills, as well daily oil runoff from urbanized areas can lead to polycyclic aromatic hydrocarbon (PAHs) entering marine ecosystems. Once PAHs enter the marine environment, fish can be exposed to them via ingestion, ventilation of the gills, and dermal uptake. The major route of uptake will depend on the behavior of the species of fish and the physicochemical properties of the PAH of concern. Habitat can be a major deciding factor for the route of exposure. For example, demersal fish or fish that consume demersal fish are highly likely to ingest PAHs that have sorbed to the sediment, whereas fish that swim at the surface are at a higher risk for dermal exposure. Upon coming in contact with a PAH, bioavailability will effect how readily the PAH is taken up. The EPA identifies 16 major PAHs of concern and each of these PAHs has a different degree of bioavailability. For instance, PAHs with lower molecular weight are more bioavailable because they dissolve more readily in water and are therefore more bioavailable for fish within the water column. Similarly, hydrophilic PAHs are more bioavailable for uptake by fish. For this reason, usage of oil dispersants, like Corexit, to treat oil spills can increase the uptake of PAHs by increasing their solubility in water and making them more available for uptake via the gills. Once a PAH is taken up, the fish's metabolism can affect the duration and intensity of the exposure to target tissues. Fish are able to readily metabolize 99% of PAHs to a more hydrophilic metabolite through their hepato-bilary system. This allows for the excretion of PAHs. The rate of metabolism of PAHs will depend on the sex and size of the species. The ability to metabolize PAHs into a more hydrophillic form can prevent bioaccumulation and halt PAHs from being passed on to organisms further up the food web. Because oil can persist in the environment long after oil spills via sedimentation, demersal fish are likely to be continually exposed to PAHs many years after oil spills. This has been proven by looking at the biliary PAH metabolites of bottom dwelling fish. For instance, bottom dwelling fish still showed elevated levels of low molecular weight PAH metabolites 10 years after *Exxon Valdez* oil spill.

Crude Oil Components

Crude oil is composed of more than 17,000 compounds. Among these 17,000 compounds are PAHs, which are considered the most toxic components of oil. PAHs are formed by pyrogenic and petrogenic processes. Petrogenic PAHs are formed by the elevated pressure of organic material. In contrast, pyrogenic PAHs are formed through the incomplete combustion of organic material. Crude oil naturally contains petrogenic PAHs and these PAH levels are increased significantly through the burning of oil which creates pyrogenic PAHs. The level of PAHs found in crude oil differs with the type of crude oil. For example, crude oil from the *Exxon Valdez* oil spill had PAH concentrations of 1.47%, while PAH concentrations from the North Sea have much lower PAH concentrations of 0.83%.

Sources of Crude Oil Pollution

Crude oil contamination in marine ecosystems can lead to both pyrogenic and petrogenic PAHs entering these ecosystems. Petrogenic PAHs can enter waterways through oil seeps, major oil spills, creosote and fuel oil runoff from urban areas. Pyrogenic PAH sources consist of diesel soot tire rubber and coal dust. Although there are natural sources of PAHs such as volcanic activity and seepage of coal deposits, anthropogenic sources pose the most significant input of PAHs into the environment. These anthroprogenic sources include residential heating, asphalt production, coal gasification, and petroleum usage. Petrogenic PAH contamination is more common from crude oil spills such as *Exxon Valdez*, or oil seeps; however, with runoff pyrogenic PAHs can also be prevalent. Although major oil spills such as *Exxon Valdez* can introduce a large amount of crude oil to a localized area in a short time span, daily runoff comprises most of the oil pollution to marine ecosystems. Atmospheric deposition can also be a source of PAHs into marine ecosystems. The deposition of PAHs from the atmosphere into a water body is largely influenced by the gas-particle partitioning of the PAH.

Effects

Many effects of PAH exposure have been observed in marine fish. Specifically, studies have been conducted on the embryonic and larval fish, the development of fish exposed to PAHs, and uptake of PAHs by fish via various routes of exposure. One study on found that Pacific herring eggs exposed to conditions mimicking the "Exxon Valdez" oil spill resulted in premature hatching of eggs, reduced size as fish matured and significant teratogenic effects, including skeletal, cardiovascular, fin and yolk sac malformations. Yolk sac edema was responsible for the majority of herring larval mortality. The teratogenic malformations in the dorsal fin and spine, and in the jaw were observed to effectively decrease the survival of developing fish, through the impairing of swimming and feeding ability respectively. Feeding and prey avoidance via swimming are crucial for the survival of larval and juvenile fish. All effects observed in herring eggs in the study were consistent with effects observed in exposed fish eggs following the *Exxon Valdez* oil spill. Zebrafish embryos exposed to oil were observed to have severe teratogenic defects similar to those seen in herring embryos, including edema, cardiac dysfunction, and intracranial hemorrhages. In a study focused on the uptake of PAHs by fish, salmon embryos were exposed to crude oil in three various situations, including via effluent from oil coated gravel. PAH concentrations in embryos directly exposed to oil and those exposed to PAH effluent were not significantly different. PAH exposure was observed to lead to death, even when the PAHs were exposed to fish via effluent. From the results, it was determined that fish embryos near the *Exxon Valdez* spill in Prince William Sound that were not directly in contact with oil still may have accumulated lethal levels of PAHs. While many laboratory and natural studies have observed significant adverse effects of PAH exposure to fish, a lack of effects has also been observed for certain PAH compounds, which could be due to a lack of uptake during exposure to the compound.

Proposed Mechanism of Toxic Action

While it has been proven that different classes of PAHs act through distinct toxic mechanisms due to the variations in their molecular weight, ring arrangements, and water solubility properties, the specific mechanisms of PAH toxicity to fish and fish development are still unknown. The proposed mechanisms of toxicity of PAHs are toxicity through narcosis, interaction with the AhR pathway, alkyl phenanthrene toxicity, and additive toxicity by multiple mechanisms.

- The narcosis model was not able to accurately predict the outcome of PAH mixture exposure of herring and pink salmon, according to a study.

- The primary toxicity of these PAHs in fish embryos has been observed to be AhR independent, and their cardiac effects are not associated with AhR activation or Cytochrome P450, family 1, member A induction in the endocardium.

- The alkyl phenanthrene model has been studied by exposing herring and pink salmon to mixtures of PAHs in an attempt to better understand the toxicity mechanisms of PAHs. The model was found to generally predict the outcomes of sublethal and lethal exposures. Oxidative stress and effects on cardiovascular morphogenesis are proposed mechanisms for alkyl phenanthrene toxicity. The specific pathway is unknown.

- Since PAHs contain many different variations of PAHs, the toxicity may be explained by using multiple mechanisms of action.

Overfishing

400 tons of jack mackerel caught by a Chilean purse seiner

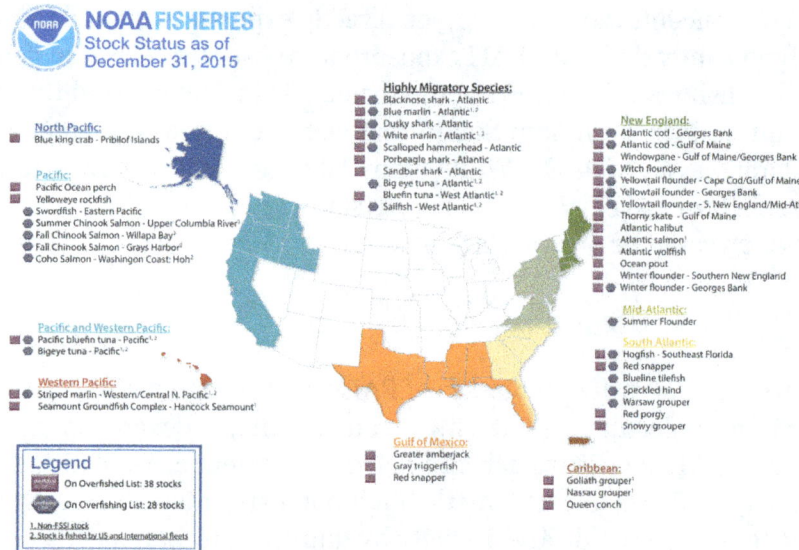

Overfished US stocks, 2015

Overfishing is a form of overexploitation where fish stocks are reduced to below acceptable levels. Overfishing can occur in water bodies of any sizes, such as ponds, rivers, lakes or oceans, and can result in resource depletion, reduced biological growth rates and low biomass levels. Sustained overfishing can lead to critical depensation, where the fish population is no longer able to sustain itself. Some forms of overfishing, for example the overfishing of sharks, has led to the upset of entire marine ecosystems.

The ability of a fishery to recover from overfishing depends on whether the ecosystem's conditions are suitable for the recovery. Dramatic changes in species composition can result in an ecosystem shift, where other equilibrium energy flows involve species compositions different from those that had been present before the depletion of the original fish stock. For example, once trout have been overfished, carp might take over in a way that makes it impossible for the trout to re-establish a breeding population.

Overfishing occurs when more fish are caught than the population can replace through natural reproduction. Gathering as many fish as possible may seem like a profitable practice, but overfishing has serious consequences. The results not only affect the balance of life in the oceans, but also the social and economic well-being of the coastal communities who depend on fish for their way of life.

Global Overfishing

Overfishing has greatly affected many fisheries around the world. As much as 85% of the world's fisheries may be over-exploited, depleted, fully exploited or in recovery from exploitation. Significant overfishing has been observed in pre-industrial times. In particular, the overfishing of the western Atlantic Ocean from the earliest days of European colonisation of the Americas has been well documented. Following World War Two, industrial fishing rapidly expanded with rapid increases in worldwide fishing catches. However, many fisheries have either collapsed or degraded to a point where increased catches are no longer possible.

Daniel Pauly, a fisheries scientist known for pioneering work on the human impacts on global fisheries, has commented:

It is almost as though we use our military to fight the animals in the ocean. We are gradually winning this war to exterminate them. And to see this destruction happen, for nothing really – for no reason – that is a bit frustrating. Strangely enough, these effects are all reversible, all the animals that have disappeared would reappear, all the animals that were small would grow, all the relationships that you can't see any more would re-establish themselves, and the system would re-emerge.

Examples of Overfishing

Examples of overfishing exist in areas such as the North Sea, the Grand Banks of Newfoundland and the East China Sea. In these locations, overfishing has not only proved disastrous to fish stocks but also to the fishing communities relying on the harvest. Like other extractive industries such as forestry and hunting, fisheries are susceptible to economic interaction between ownership or stewardship and sustainability, otherwise known as the tragedy of the commons.

- The Peruvian coastal anchovy fisheries crashed in the 1970s after overfishing and an El Niño season largely depleted anchovies from its waters. Anchovies were a major natural resource in Peru; indeed, 1971 alone yielded 10.2 million metric tons of anchovies. However, the following five years saw the Peruvian fleet's catch amount to only about 4 million tons. This was a major loss to Peru's economy.

- The collapse of the cod fishery off Newfoundland, and the 1992 decision by Canada to impose an indefinite moratorium on the Grand Banks, is a dramatic example of the consequences of overfishing.

- The sole fisheries in the Irish Sea, the west English Channel, and other locations have become overfished to the point of virtual collapse, according to the UK government's official Biodiversity Action Plan. The United Kingdom has created elements within this plan to attempt to restore this fishery, but the expanding global human population and the expanding demand for fish has reached a point where demand for food threatens the stability of these fisheries, if not the species' survival.

- Many deep sea fish are at risk, such as orange roughy, Patagonian toothfish, and sablefish. The deep sea is almost completely dark, near freezing and has little food. Deep sea fish grow slowly because of limited food, have slow metabolisms, low reproductive rates, and many don't reach breeding maturity for 30 to 40 years. A fillet of orange roughy at the store is probably at least 50 years old. Most deep sea fish are in international waters, where there are no legal protections. Most of these fish are caught by deep trawlers near seamounts, where they congregate because of food. Flash freezing allows the trawlers to work for days at a time, and modern fishfinders target the fish with ease.

- Blue walleye became extinct in the Great Lakes in the 1980s. Until the middle of the 20th century, it was a commercially valuable fish, with about a half million tonnes being landed during the period from about 1880 to the late 1950s, when the populations collapsed, apparently through a combination of overfishing, anthropogenic eutrophication, and competition with the introduced rainbow smelt.

- The World Wildlife Fund and the Zoological Society of London jointly issued their "Living Blue Planet Report" on 16 September 2015 which states that there was a dramatic fall of

74% in worldwide stocks of the important scombridae fish such as mackerel, tuna and bonitos between 1970 and 2010, and the global overall "population sizes of mammals, birds, reptiles, amphibians and fish fell by half on average in just 40 years."

Examples of Good Fisheries Management

Several countries are now effectively managing their fisheries. Examples include Iceland and New Zealand. The United States has turned many of its fisheries around from being in a highly depleted state.

Consequences

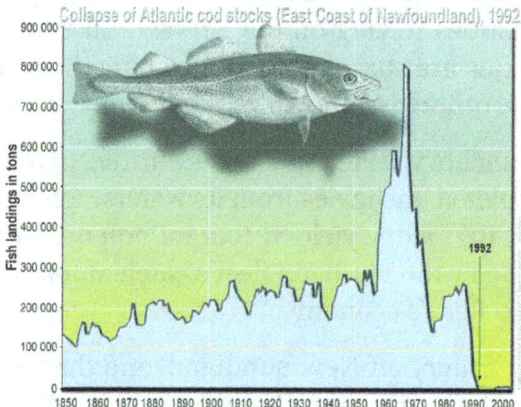

Atlantic cod stocks were severely overfished in the 1970s and 1980s, leading to their abrupt collapse in 1992

According to a 2008 UN report, the world's fishing fleets are losing US$50 billion each year through depleted stocks and poor fisheries management. The report, produced jointly by the World Bank and the UN Food and Agriculture Organization (FAO), asserts that half the world's fishing fleet could be scrapped with no change in catch. In addition, the biomass of global fish stocks have been allowed to run down to the point where it is no longer possible to catch the amount of fish that could be caught. Increased incidence of schistosomiasis in Africa has been linked to declines of fish species that eat the snails carrying the disease-causing parasites. Massive growth of jellyfish populations threaten fish stocks, as they compete with fish for food, eat fish eggs, and poison or swarm fish, and can survive in oxygen depleted environments where fish cannot; they wreak massive havoc on commercial fisheries. Overfishing eliminates a major jellyfish competitor and predator exacerbating the jellyfish population explosion. Both climate change and a restructuring of the ecosystem have been found to be major roles in an increase in jellyfish population in the Irish Sea in the 1990s.

Types

There are three recognized types of biological overfishing: growth overfishing, recruit overfishing and ecosystem overfishing.

Growth Overfishing

Growth overfishing occurs when fish are harvested at an average size that is smaller than the size that would produce the maximum yield per recruit. A recruit is an individual that makes it to maturity, or into the limits specified by a fishery, which are usually size or age. This makes the total yield less than it would be if the fish were allowed to grow to an appropriate size. It can be countered by reducing fishing mortality to lower levels and increasing the average size of harvested fish to a size that will allow maximum yield per recruit.

Recruitment Overfishing

Recruitment overfishing occurs when the mature adult population (spawning biomass) is depleted to a level where it no longer has the reproductive capacity to replenish itself—there are not enough adults to produce offspring. Increasing the spawning stock biomass to a target level is the approach taken by managers to restore an overfished population to sustainable levels. This is generally accomplished by placing moratoriums, quotas and minimum size limits on a fish population.

Ecosystem Overfishing

Ecosystem overfishing occurs when the balance of the ecosystem is altered by overfishing. With declines in the abundance of large predatory species, the abundance of small forage type increases causing a shift in the balance of the ecosystem towards smaller fish species.

Acceptable Levels

The notion of overfishing hinges on what is meant by an acceptable level of fishing. More precise biological and bioeconomic terms define acceptable level as follows:

- Biological overfishing occurs when fishing mortality has reached a level where the stock biomass has negative marginal growth (reduced rate of biomass growth), as indicated by the red area in the figure. (Fish are being taken out of the water so quickly that the replenishment of stock by breeding slows down. If the replenishment continues to diminish for long enough, replenishment will go into reverse and the population will decrease).

- Economic or bioeconomic overfishing additionally considers the cost of fishing when determining acceptable catches. Under this framework, a fishery is considered to be overfished when catches exceed maximum economic yield where resource rent is at its maximum. Fish are being removed from the fishery so quickly that the profitability of the fishery is sub-optimal. A more dynamic definition of economic overfishing also considers the present value of the fishery using a relevant discount rate to maximise the flow of resource rent over all future catches.

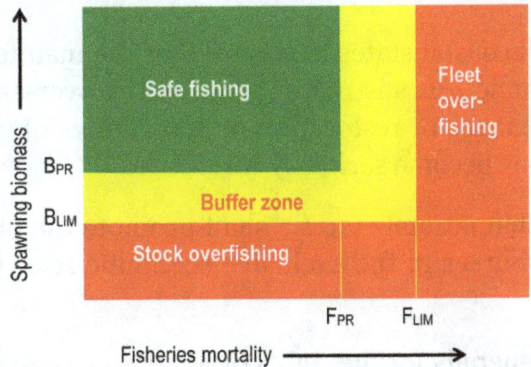

The Traffic Light colour convention, showing the concept of Harvest Control Rule (HCR), specifying when a rebuilding plan is mandatory in terms of precautionary and limit reference points for spawning biomass and fishing mortality rate.

Harvest Control Rule

A model proposed in 2010 for predicting acceptable levels of fishing is the Harvest Control Rule (HCR), which is a set of tools and protocols with which management has some direct control of harvest rates and strategies in relation to predicting stock status, and long-term maximum sustainable yields. Constant catch and constant fishing mortality are two types of simple harvest control rules.

Input and Output Orientations

Fishing capacity can also be defined using an input or output orientation.

- An input-oriented fishing capacity is defined as the maximum available capital stock in a fishery that is fully utilized at the maximum technical efficiency in a given time period, given resource and market conditions.

- An output-oriented fishing capacity is defined as the maximum catch a vessel (fleet) can produce if inputs are fully utilized given the biomass, the fixed inputs, the age structure of the fish stock, and the present stage of technology.

Technical efficiency of each vessel of the fleet is assumed necessary to attain this maximum catch. The degree of capacity utilization results from the comparison of the actual level of output (input) and the capacity output (input) of a vessel or a fleet.

Mitigation

With present and forecast world population levels it is not possible to solve the over fishing issue; however, there are mitigation measures that can save selected fisheries and forestall the collapse of others.

In order to meet the problems of overfishing, a precautionary approach and Harvest Control Rule (HCR) management principles have been introduced in the main fisheries around the world. The Traffic Light color convention introduces sets of rules based on predefined critical values, which could be adjusted as more information is gained.

The United Nations Convention on the Law of the Sea treaty deals with aspects of over fishing in articles 61, 62, and 65.

- Article 61 requires all coastal states to ensure that the maintenance of living resources in their exclusive economic zones is not endangered by over-exploitation. The same article addresses the maintenance or restoration of populations of species above levels at which their reproduction may become seriously threatened.

- Article 62 provides that coastal states: "shall promote the objective of optimum utilization of the living resources in the exclusive economic zone without prejudice to Article 61".

- Article 65 provides generally for the rights of, inter alia, coastal states to prohibit, limit, or regulate the exploitation of marine mammals.

According to some observers, overfishing can be viewed as an example of the tragedy of the commons; appropriate solutions would therefore promote property rights through, for instance, privatization and fish farming. Daniel K. Benjamin, in *Fisheries are Classic Example of the "Tragedy of the Commons"*, cites research by Grafton, Squires and Fox to support the idea that privatization can solve the overfishing problem:

> *According to recent research on the British Columbia halibut fishery, where the commons has been at least partly privatized, substantial ecological and economic benefits have resulted. There is less damage to fish stocks, the fishing is safer, and fewer resources are needed to achieve a given harvest.*

Another possible solution, at least for some areas, is quotas, so fishermen can only legally take a certain amount of fish. A more radical possibility is declaring certain areas of the sea "no-go zones" and make fishing there strictly illegal, so the fish in that area have time to recover and repopulate.

Controlling consumer behavior and demand is a key in mitigating action. Worldwide, a number of initiatives emerged to provide consumers with information regarding the conservation status of the seafood available to them. The Guide to Good Fish Guides lists a number of these.

Government Regulation

Many regulatory measures are available for controlling overfishing. These measures include fishing quotas, bag limits, licensing, closed seasons, size limits and the creation of marine reserves and other marine protected areas.

A model of the interaction between fish and fishers showed that when an area is closed to fishers, but there are no catch regulations such as individual transferable quotas, fish catches are temporarily increased but overall fish biomass is reduced, resulting in the opposite outcome from the one desired for fisheries. Thus, a displacement of the fleet from one locality to another will generally have little effect if the same quota is taken. As a result, management measures such as temporary closures or establishing a marine protected area of fishing areas are ineffective when not combined with individual fishing quotas. An inherent problem with quotas is that fish populations vary from year to year. A study has found that fish populations rise dramatically after stormy years due to more nutrients reaching the surface and therefore greater primary production. To fish sustainably, quotas need to be changed each year to account for fish population.

Individual transferable quotas (ITQs) are fishery rationalization instruments defined under the Magnuson-Stevens Fishery Conservation and Management Act as limited access permits to harvest quantities of fish. Fisheries scientists decide the optimal amount of fish (total allowable catch) to be harvested in a certain fishery. The decision considers carrying capacity, regeneration rates and future values. Under ITQs, members of a fishery are granted rights to a percentage of the total allowable catch that can be harvested each year. These quotas can be fished, bought, sold, or leased allowing for the least cost vessels to be used. ITQs are used in New Zealand, Australia, Iceland, Canada, and the United States. Only three ITQ programs have been implemented in the United States due to a moratorium supported by Ted Stevens.

In 2008, a large-scale study of fisheries that used ITQs compared to ones that didn't provided strong evidence that ITQs can help to prevent collapses and restore fisheries that appear to be in decline.

China bans fishing in the South China Sea for a period each year.

Removal of Subsidies

Several scientists have called for an end to subsidies paid to deep sea fisheries. In international waters beyond the 200 nautical mile exclusive economic zones of coastal countries, many fisheries are unregulated, and fishing fleets plunder the depths with state-of-the-art technology. In a few hours, massive nets weighing up to 15 tons, dragged along the bottom by deep-water trawlers, can destroy deep-sea corals and sponge beds that have taken centuries or millennia to grow. The trawlers can target orange roughy, grenadiers, or sharks. These fish are usually long-lived and late maturing, and their populations take decades, even centuries to recover.

Fisheries scientist Daniel Pauly and economist Ussif Rashid Sumaila have examined subsidies paid to bottom trawl fleets around the world. They found that US$152 million per year are paid to deep-sea fisheries. Without these subsidies, global deep-sea fisheries would operate at a loss of $50 million a year. A great deal of the subsidies paid to deep-sea trawlers is to subsidize the large amount of fuel required to travel beyond the 200-mile limit and drag weighted nets.

> "There is surely a better way for governments to spend money than by paying subsidies to a fleet that burns 1.1 billion litres of fuel annually to maintain paltry catches of old growth fish from highly vulnerable stocks, while destroying their habitat in the process" – *Pauly*.

> "Eliminating global subsidies would render these fleets economically unviable and would relieve tremendous pressure on over-fishing and vulnerable deep-sea ecosystems" – *Sumaila*.

Minimizing Fishing Impact

Fishing techniques may be altered to minimize bycatch and reduce impacts on marine habitats. These techniques include using varied gear types depending on target species and habitat type. For example, a net with larger holes will allow undersized fish to avoid capture. A turtle excluder device (TED) allows sea turtles and other megafauna to escape from shrimp trawls. Avoiding fishing in spawning grounds may allow fish stocks to rebuild by giving adults a chance to reproduce.

Aquaculture

Aquaculture involves the farming of fish in captivity. This approach effectively privatizes fish stocks and creates incentives for farmers to conserve their stocks. It also reduces environmental impact. However, farming carnivorous fish, such as salmon, does not always reduce pressure on wild fisheries, since carnivorous farmed fish are usually fed fishmeal and fish oil extracted from wild forage fish.

Aquaculture played a minor role in the harvesting of marine organisms until the 1970s. Growth in aquaculture increased rapidly in 1990s when the rate of wild capture plateaued. Aquaculture now provides approximately half of all harvested aquatic organisms. Aquaculture production rates continue to grow while wild harvest remains steady.

regime of fishing resources. Overexploitation and rent dissipation of fishermen arise in open-access fisheries as was shown in Gordon.

In open-access resources like fish stocks, in the absence of a system like individual transferable quotas, the impossibility of excluding others provokes the fishermen who want to increase catch to do so effectively by taking someone else' share, intensifying competition. This tragedy of the commons provokes a capitalization process that leads them to increase their costs until they are equal to their revenue, dissipating their rent completely.

Resistance from Fishermen

> There is always disagreement between fishermen and government scientists... Imagine an overfished area of the sea in the shape of a hockey field with nets at either end. The few fish left therein would gather around the goals because fish like structured habitats. Scientists would survey the entire field, make lots of unsuccessful hauls, and conclude that it contains few fish. The fishermen would make a beeline to the goals, catch the fish around them, and say the scientists do not know what they are talking about. The subjective impression the fishermen get is always that there's lots of fish - because they only go to places that still have them... fisheries scientists survey and compare entire areas, not only the productive fishing spots. – *Fisheries scientist Daniel Pauly*

Over Fishing and Mitigation

Overfishing can be defined in a number of ways. However, everything comes down to one simple point: Catching too much fish for the system to support leads to an overall degradation to the system. Overfishing is a non-sustainable use of the oceans.

Below are a few definitions in use by organizations and governments.

The practice of commercial and non-commercial fishing which depletes a fishery by catching so many adult fish that not enough remain to breed and replenish the population. Overfishing exceeds the carrying capacity of a fishery.

Catching too many fish; fishing so much that the fish cannot sustain their population. The fish get fewer and fewer, until finally there are none to catch.

Fishing with a sufficiently high intensity to reduce the breeding stock levels to such an extent that they will no longer support a sufficient quantity of fish for sport or commercial harvest.

Worldwide, fishing fleets are two to three times as large as needed to take present day catches of fish and other marine species and as what our oceans can sustainably support. On a global scale we have enough fishing capacity to cover at least four Earth like planets.

On top of the overcapacity many fishing methods are unsustainable in their own way. These methods have a large impact on the basic functioning of our marine ecosystems. These unselective fishing practices and gear cause tremendous destruction on non target species. By catch discards and bottom trawling destruction are two examples of this.

Overfishing occurs when fishing activities reduce fish stocks below an acceptable level. This can occur in any body of water from a pond to the oceans. The notion of overfishing hinges on what is meant by an acceptable level of fishing. More precise biological and bioeconomic terms define acceptable level as follows:

Biological overfishing occurs when fishing mortality has reached a level where the stock biomass has negative marginal growth (slowing down biomass growth), as indicated by the red area in the figure. (Fish are being taken out of the water so quickly that the replenishment of stock by breeding slows down. If the replenishment continues to slow down for long enough, replenishment will go into reverse and the population will decrease.)

Economic or bioeconomic overfishing additionally considers the cost of fishing and defines overfishing as a situation of negative marginal growth of resource rent. (Fish are being taken out of the water so quickly that the growth in the profitability of fishing slows down. If this continues for long enough, profitability will decrease.) Resource rent is an economic term of abnormal or supernormal profit which derives from the exploitation of natural resources. A more dynamic definition of economic overfishing may also include a relevant discount rate and present value of flow of resource rent over all future catches.Ultimately overfishing may lead to resource depletion in cases of subsidized fishing, low biological growth rates and critical low biomass levels. Particularly, overfishing of sharks has led to the upset of entire marine ecosystems.

The ability of the fisheries to naturally recover also depends on whether the conditions of the ecosystems are suitable for population growth. Dramatic changes in species composition may establish other equilibrium energy flows that involve other species compositions than had been present before (ecosystem shift). For example, remove nearly all the trout and the carp might take over and make it nearly impossible for the trout to re-establish a breeding population.

Types of overfishing:

There are three recognized types of overfishing:

1. Growth overfishing

2. Recruit overfishing

3. Ecosystem overfishing

Growth overfishing is when fishes are harvested at an average size that is smaller than the size that would produce the maximum yield per recruit. Thus making the total yield less than it would be if the fish were allowed to grow to a reasonable size. Reducing fishing mortality to lower levels and increasing the average size of the fishes harvested to a length that would allow maximum yield per recruit.

Recruit overfishing is when the mature adult (spawning biomass) population is depleted to a level where it no longer has the reproductive capacity to replenish itself. There are not enough adults to produce offspring. Increasing the spawning stock biomass to a target level is the approach taken by managers to restore an overfished population to sustainable levels. This is generally accomplished by placing moratoriums, quotas and minimum size limits on a fish population.

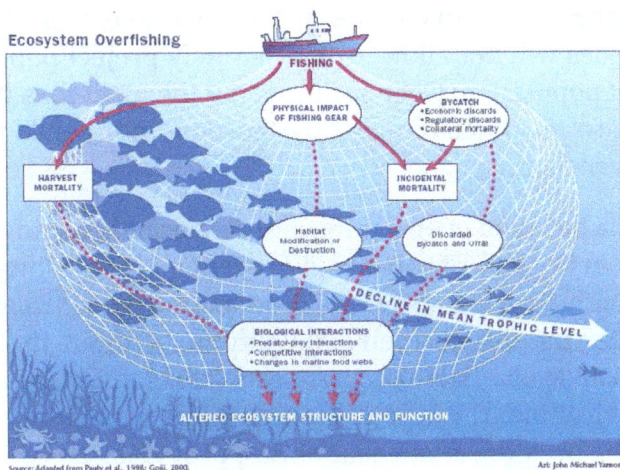

Ecosystem overfishing is when the balance of the ecosystem is altered due to overfishing. Declines in the abundances of large predatory species declines and in turn small forage type species increase in abundance, causing a shift in the balance of the ecosystem towards smaller species of fish.

Instances of Overfishing

The number of endangered fishes in the United States continues to grow at a frightening rate. From the time of the first Anglo-European settlement in the New World, one-fourth of all fish species have become imperiled or gone extinct. The same holds true for one-half of freshwater mussel species, one-third of native crayfish, and one-fourth of all amphibian species.

In many cases their numbers have been driven below critical levels by over-fishing. This has occurred to the Atlantic cod of the North Atlantic, the salmon of the Pacific Northwest, and to many species in estuaries near cities, such as rockfish and lingcod. Traditionally, fish populations were believed to be almost fantastically resilient. In the beginning, there were so many fish that people had trouble believing they could ever make a dent in their populations. A good example of this was the Pacific salmon, which at one time provided an abundance of cheap and tasty protein to residents of the West Coast. There are tales of salmon as big as a man, and so many swimming at once in a stream that a person could walk across their backs to the other bank. Atlantic cod, too, provided livings for fishing families in America for over two hundred years.

Fish Populations

This idea of resilient populations was in many ways made worse by population research techniques which mathematically indicated that the harder a population was fished, the more fishes the system would produce. There were several problems with this. First, the harder a population is fished, the smaller the individuals become. This means fishermen must catch more of them, wiping out any advantage of greater numbers.

Second, there is a critical level below which the population must not fall, else it will likely be doomed to extinction. Over-fishing, to succeed, depends on total cooperation from commercial interests as well as intimate knowledge of population numbers by scientists. We have learned, tragically, that only rarely will commercial interests cooperate with management measures. Basically, the race is on to scoop up as many fishes as possible in the shortest period of time. Worse, we've learned that our best scientific techniques will not allow us to accurately project populations, or even to accurately determine a current population. Over-fishing is not the great idea it used to be--and without the ability to determine current populations, it has become impossible to draw the line between sustainable fishing and over-fishing.

Finally, even when the line between over-fishing and sustainable fishing is accurately drawn, the whole concept falls apart in the face of catastrophic events. When drought strikes, or a chemical spill, or unseasonable weather, or a failure of a food source, or a disease, fishes need the extra numbers in order to buffer the loss to their populations. Without the extra numbers, populations sink below the critical levels. In short, there was a reason for the abundance of animals in a healthy, undamaged environment. Without enough animals, populations can go extinct during even the most minor of the periodic crises that nature provides.

Mitigation: Trying to Replace What's Gone

Mitigation is a popular strategy for managing endangered fish populations. Mitigation is the process of trying to replace something that has been lost. When a commercial interest plans to destroy an environment, for example to drain a marsh, build a dam, or build close enough to a water body that its health will be endangered, environmental protection agencies are called in to evaluate the situation and use federal and state law to assist endangered species.

Mitigation has been presented in these laws as a viable solution to any kind of development. Therefore, in only rare cases is a commercial interest required to modify its plans in any way, and almost never to cancel them. Instead, it spends money on mitigation. For example, a company may drain and pave over a marsh that served as habitat for endangered species and a resting stop for migrating birds, and replace it with an artificial pond ten miles away. In reality, even when such effort and expense is made that the new artificial environment looks just like the old to an inexperienced eye, it is nevertheless a sterile, lifeless environment that will be incapable of supporting the wildlife of the environment that it replaces for many decades or even centuries.

Legislation now requires rehabilitation of certain areas, e.g. creation of wetland. Wetland (or other habitat) would be created to —mitigate‖ or offset (compensate for) losses elsewhere. Mitigation is really restoration in where the targeted habitat never existed. Compensatory mitigation involves the creation of a habitat of equal size/value in trade for the loss/destruction of existing habitat. In most cases, mitigation requires creating more habitat (2 to 1, 3 to 1) than what is lost to development. The problem is the certain destruction of a vanishing habitat for the uncertain promise of similar habitat, very contentious issue.

Problems with Mitigation:

- Often the local flow regime or sediment types are inappropriate for creating the new habitat

- To create a new habitat, the existing habitat, desirable or not, is destroyed

- Mitigation can proceed at a site some distance from the damage

With present and forecast levels of the world population it is not possible to solve the overfishing issue; however, there are mitigation measures that can save selected fisheries and forestall the collapse of others.

In order to meet the problems of overfishing, a precautionary approach and Harvest Control Rule (HCR) management principles have been introduced in the main fisheries around the world. The Traffic Light colour convention introduces sets of rules based on predefined critical values, which could be adjusted as more information is gained.

Habitat conservation

- Our tendency to be reactive instead of proactive is self-destructive and is contrary to the essence of life: to maintain our existence.

- Both ocean and land conservation is important.

- Preserving habitats is essential to preserving biodiversity.

- Establishing protected areas.

    Zone Reserve Model

    Marine Protected Areas and Marine Reserves

    ex: NOAA Marine Sanctuary

    USA Marine Reserves in New Zealand

NOAA's Center for Coastal Fisheries and Habitat Research

A mission was undertaken which provides coastal managers with scientific information and tools to make informed stewardship decisions about one of NOAA's protected areas—the Tortugas North Ecological Reserve. The Reserve was designated in July 2001 to protect and preserve the diverse marine life and special habitats in critical areas of the Florida Keys National Marine Sanctuary. In doing this, NOAA strives to balance society's environmental, social, and economic goals.

The Reserve, located west of the Dry Tortugas National Park, has an area of 90 square nautical miles. It includes the coral reefs of the Tortugas Banks and the soft-bottom habitats of the West Florida Shelf. An ecological reserve is one type of marine zoning intended to protect habitats and the species using those habitats – an element of marine spatial planning.

Monterey Bay National Marine Sanctuary

Monterey Bay National Marine Sanctuary is the largest national marine sanctuary and one of the largest marine protected areas in the United States. Within the boundaries of the sanctuary is a rich array of habitats, from rugged rocky shores and lush kelp forests to one of the largest underwater canyons in North America. These habitats abound with life, from tiny microscopic plants to enormous blue whales. The sanctuary is home to a diversity of species including marine mammals, seabirds and shorebirds, sea turtles, fishes, invertebrates, and marine algae.

Congressionally designated in 1992 as a National Marine Sanctuary for the purpose of resource protection, research, education, and public use.

- Includes bays, estuaries, coastal and oceanic waters

- High diversity of flora and fauna including 33 species of marine mammals, 94 species of seabird, 345 species of fishes, and numerous species of invertebrates and plants

- Contains the Monterey Canyon, a submarine canyon that rivals the Grand Canyon in size

- Contains an estimated 225 documented shipwrecks or lost aircraft and 718 historic sites

A new management plan for Monterey Bay National Marine Sanctuary was released in November 2008, and it contains a number of management actions that will address current issues and concerns. The plan stresses an ecosystem-based approach to management, which requires consideration of ecological interrelationships not only within the sanctuary, but within the larger context of the California Current ecosystem. It also makes essential an increased level of cooperation with other management agencies in the region. The management plan includes twenty-nine action plans that will guide the sanctuary for the next five to ten years.

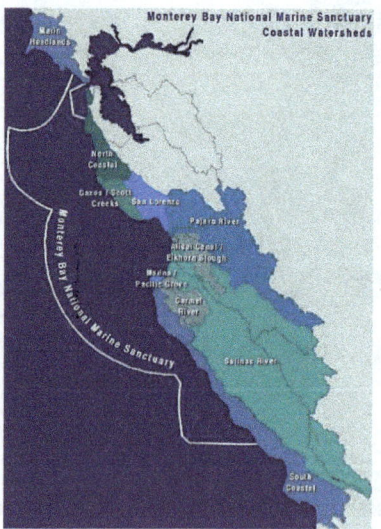

Channel Islands National Marine Sanctuary

Channel Islands National Marine Sanctuary contains spectacularly rich and diverse marine life. With a variety of habitats including kelp forests, sandy bottom, and open ocean, it is home to diverse fish and invertebrate communities, serves as part of the migratory route of whales, and as feeding and breeding grounds for seabirds and marine mammals. Located offshore of Southern California, the sanctuary is adjacent to the growing counties of Ventura and Santa Barbara, and not far from the heavily populated Los Angeles metropolitan area, bringing to it a variety of recreational and commercial human activities, including diving, kayaking, fishing, boating, wildlife viewing, and shipping.

- The Chumash were the first people to inhabit the Channel Islands.

- The islands were first visited by Europeans in 1542.

- In the 1800s the islands served as a location for sea otter, seal, and sea lion hunting. Subsequently, the land was cultivated for ranching and farming purposes.

- The sanctuary was designated on Sept. 22, 1980, and encompasses 1,470 square statute miles (1,110 square nautical miles).

- In 2003, 12 marine protected areas were designated by the California Department of Fish and Game Commission.

- In 2007 several of the marine protected areas were extended to the federal boundary and one new area was created.

- Numerous shipwrecks are located in waters surrounding the islands.

The sanctuary is an important area for recreational and commercial use, including diving, kayaking, fishing, boating, wildlife viewing, shipping transit, and research.

The sanctuary contains a network of marine zones established in state waters in 2003 and extended to the federal boundary in 2007 that will help protect these valuable resources. These marine zones now include 11 no-take zones (also called marine reserves) and two marine conservation areas where some fishing is allowed. In addition, a new management plan for the Channel Islands sanctuary was released in 2009; it recommends a number of management actions that will address concerns of resource protection and management. The plan stresses an ecosystem-based approach to management that requires consideration of ecological interrelationships not only within the sanctuary, but within the larger context of the Santa Barbara Channel. Specific management recommendations include an improved water quality monitoring program, actions to reduce vessel discharges, and directed research on emerging issues.

Overfishing can be viewed as a case of the tragedy of the commons; in that sense, solutions would promote property rights, such as privatization and fish farming.

According to recent research on the British Columbia halibut fishery, where the commons has been at least partly privatized, substantial ecological and economic benefits have resulted. There is less damage to fish stocks, the fishing is safer, and fewer resources are needed to achieve a given harvest.

Another possible solution, at least for some areas, is fishing quotas, so fishermen can only legally take a certain amount of fish. A more radical possibility is declaring certain areas of the sea —no- go zones‖ and make fishing there strictly illegal, so the fish in that area have time to recover and repopulate.

Controlling consumer behaviour and demand is a key in mitigation action. Worldwide a number of initiatives emerged to provide consumers with information regarding the conservation status of the seafood available to them.

Bycatch

Bycatch, in the fishing industry, is a fish or other marine species that is caught unintentionally while catching certain target species and target sizes of fish, crabs etc. Bycatch is either of a different species, the wrong sex, or is undersized or juvenile individuals of the target species. The term "bycatch" is also sometimes used for untargeted catch in other forms of animal harvesting or collecting.

Shrimp bycatch

In 1997, the Organisation for Economic Co-operation and Development (OECD) defined bycatch as "total fishing mortality, excluding that accounted directly by the retained catch of target species". Bycatch contributes to fishery decline and is a mechanism of overfishing for unintentional catch.

The fisherman bycatch issue originated due to the "mortality of dolphins in tuna nets in the 1960s"

There are at least four different ways the word "bycatch" is used in fisheries:

- Catch which is retained and sold but which is not the target species for the fishery

- Species/sizes/sexes of fish which fishermen discard

- Non-target fish, whether retained and sold or discarded

- Unwanted invertebrate species, such as echinoderms and non-commercial crustaceans, and various vulnerable species groups, including seabirds, sea turtles, marine mammals and elasmobranchs (sharks and their relatives).

Examples

Recreational Fishing

Given the popularity of recreational fishing throughout the world, a small local study in the US in 2013 suggested that discards may be an important unmonitored source of fish mortality.

Shrimp Trawling

Double-rigged shrimp trawler hauling in the nets

Shrimp bycatch

The highest rates of incidental catch of non-target species are associated with tropical shrimp trawling. In 1997, the Food and Agriculture Organization of the United Nations (FAO) documented the estimated bycatch and discard levels from shrimp fisheries around the world. They found discard rates (bycatch to catch ratios) as high as 20:1 with a world average of 5.7:1.

Shrimp trawl fisheries catch 2% of the world total catch of all fish by weight, but produce more than one-third of the world total bycatch. American shrimp trawlers produce bycatch ratios between 3:1 (3 bycatch:1 shrimp) and 15:1(15 bycatch:1 shrimp).

Trawl nets in general, and shrimp trawls in particular, have been identified as sources of mortality for cetacean and finfish species. When bycatch is discarded (returned to the sea), it is often dead or dying.

Tropical shrimp trawlers often make trips of several months without coming to port. A typical haul may last 4 hours after which the net is pulled in. Just before it is pulled on board the net is washed by zigzagging at full speed. The contents are then dumped on deck and are sorted. An average of 5.7:1 means that for every kilogram of shrimp there are 5.7 kg of bycatch. In tropical inshore waters the bycatch usually consists of small fish. The shrimps are frozen and stored on-board; the bycatch is discarded.

Recent sampling in the South Atlantic rock shrimp fishery found 166 species of finfish, 37 crustacean species, and 29 other species of invertebrate among the bycatch in the trawls. Another sampling of the same fishery over a two-year period found that rock shrimp amounted to only 10% of total catch weight. Iridescent swimming crab, dusky flounder, inshore lizardfish, spot, brown shrimp, longspine swimming crabs, and other bycatch made up the rest.

Despite the use of bycatch reduction devices, the shrimp fishery in the Gulf of Mexico removes about 25–45 million red snapper annually as bycatch, nearly one half the amount taken in directed recreational and commercial snapper fisheries.

Cetacean

Cetaceans, such as dolphins, porpoises, and whales, can be seriously affected by entanglement in fishing nets and lines, or direct capture by hooks or in trawl nets. Cetacean bycatch is increasing in intensity and frequency. In some fisheries, cetaceans are captured as bycatch but then retained because of their value as food or bait. In this fashion, cetaceans can become a target of fisheries.

Group of Fraser's dolphins.

One example of bycatch is dolphins caught in tuna nets. As dolphins are mammals and do not have gills they may drown while stuck in nets underwater. This bycatch issue has been one of the reasons of the growing ecolabelling industry, where fish producers mark their packagings with disclaimers such as "dolphin friendly" to reassure buyers. However, "dolphin friendly" does not mean that dolphins were not killed in the production of a particular tin of tuna, but that the fleet which caught the tuna did not *specifically* target a feeding pod of dolphins, but relied on other methods to spot tuna schools.

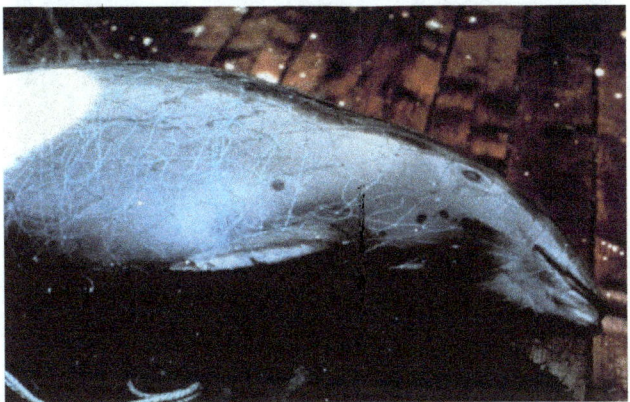

A Dall's porpoise caught in a fishing net

Albatross

This black-browed albatross has been hooked on a long-line.

Of the 21 albatross species recognised by IUCN on their Red List, 19 are threatened, and the other two are *near threatened*. Two species are considered critically endangered: the Amsterdam albatross and the Chatham albatross. One of the main threats is commercial long-line fishing, because the albatrosses and other seabirds which readily feed on offal are attracted to the set bait, become hooked on the lines and drown. An estimated 100,000 albatross per year are killed in this fashion. Unregulated pirate fisheries exacerbate the problem.

Sea Turtles

Sea turtles, already critically endangered, have been killed in large numbers in shrimp trawl nets. Estimates indicate that thousands of Kemp's ridley, loggerhead, green and leatherback sea turtles are caught in shrimp trawl fisheries in the Gulf of Mexico and the US Atlantic annually The speed and length of the trawl method is significant because, "for a tow duration of less than 10 minutes, the mortality rate for sea turtles is less than one percent, whereas for tows greater than sixty minutes the mortality rate rapidly increases to fifty to one hundred percent".

Loggerhead sea turtle

Sea turtles can sometimes escape from the trawls. In the Gulf of Mexico, the Kemp's ridley turtles recorded most interactions, followed in order by loggerhead, green, and leatherback sea turtles. In the US Atlantic, the interactions were greatest for loggerheads, followed in order by Kemp's ridley, leatherback, and green sea turtles.

Mitigation

A turtle excluder device

Concern about bycatch has led fishermen and scientists to seek ways of reducing unwanted catch. There are two main approaches.

One approach is to ban fishing in areas where bycatch is unacceptably high. Such area closures can be permanent, seasonal, or for a specific period when a bycatch problem is registered. Temporary area closures are common in some bottom-trawl fisheries where undersized fish or non-target species are caught unpredictably. In some cases fishermen are required to relocate when a bycatch problem occurs.

The other approach is alternative fishing gear. A technically simple solution is to use nets with a larger mesh size, allowing smaller species and smaller individuals to escape. However, this usually requires replacing the existing gear. In other cases, it is possible to modify gear. The "Bycatch Reduction Device" (BRD) and the Nordmore grate are net modifications that help fish escape from shrimp nets.

BRDs allow many commercial finfish species to escape. The US government has approved BRDs that reduce finfish bycatch by 30%. Spanish mackerel and weakfish bycatch in the South Atlantic was reduced by 40%. However, recent surveys suggest BRDs may be less effective than previously thought. A rock shrimp fishery off Florida found the devices did not exclude 166 species of fish, 37 crustacean species, and 29 species of other invertebrates.

In 1978, the National Marine Fisheries Service (NMFS) started to develop turtle excluder devices (TEDs). A TED uses a grid which deflects turtles and other big animals, so they exit from the trawl net through an opening above the grid. US shrimp trawlers and foreign fleets which market shrimp in the US are required to use TEDs. Not all nations enforce the use of TEDs.

For the most part, when they are used, TEDs have been successful reducing sea turtle bycatch. However, they are not completely effective, and some turtles are still captured. NMFS certifies TED designs if they are 97% effective. In heavily trawled areas, the same sea turtle may pass repeatedly through TEDs. Recent studies indicate recapture rates of twenty percent or more, but it is not clear how many turtles survive the escape process.

The size selectivity of trawl nets is controlled by the size of the net openings, especially in the "cod end". The larger the openings, the more easily small fish can escape. The development and testing of modifications to fishing gear to improve selectivity and decrease impact is called "conservation engineering."

Seabirds with longline fishing vessel

Longline fishing is controversial in some areas because of by-catch. Mitigation methods have been successfully implemented in some fisheries. These include:

- weights to sink the lines quickly

- streamer lines to scare birds away from baited hooks while deploying the lines

- setting lines only at night with minimal ship lighting (to avoid attracting birds)

- limiting fishing seasons to the southern winter (when most seabirds are not feeding young)

- not discharging offal while setting lines.

However, gear modifications do not eliminate by-catch of many species. In March 2006, the Hawai'i longline swordfish fishing season was closed due to excessive loggerhead sea turtle by-catch after being open only a few months, despite using modified circle hooks.

One solution that Norway came up with to reduce bycatch is to, "adopt a 'no discards' policy". This means that the fishermen must keep everything they catch. This policy has helped to, "encourage [bycatch] research", which, in turn has helped "encourage behavioral changes in fishers" and "reduce the waste of life" as well.

Alternative to Release

Some fisheries retain bycatch, rather than throwing the fish back into the ocean. Sometimes bycatch are sorted and sold as food, especially in Asia, Africa and Latin America where cost of labour is cheaper. Bycatch can also be sold in frozen bags as "assorted seafood" or "seafood medley" at cheaper prices. Bycatch can be converted into fish hydrolysate (ground up fish carcasses) for use as a soil amendment in organic agriculture or it can be used as an ingredient in fish meal. In Southeast Asia bycatch is sometimes used as a raw material for fish sauce production. Bycatch is also commonly de-boned, de-shelled, ground and blended into fish paste or moulded into fish cakes (surimi) and sold either fresh (for domestic use) or frozen (for export). This is commonly the case in Asia or by Asian fisheries. Sometimes bycatch are sold to fish farms to feed farmed fish, especially in Asia.

If bycatch is quickly released, predators and scavengers may consume it.

Non-fisheries Bycatch

The term "bycatch" is used also in contexts other than fisheries. Examples are insect collecting with pitfall traps or flight interception traps for either financial, controlling or scientific purposes (where the bycatch may either be small vertebrates or untargeted insects) and control of introduced vertebrates which have become pest species like the muskrat in Europe (where the bycatch in traps may be e.g. European mink or waterfowl).

Cetacean Bycatch

Cetacean bycatch is the incidental capture of non-target cetacean species such as dolphins, porpoises, and whales by fisheries. Bycatch can be caused by entanglement in fishing nets and lines, or direct capture by hooks or in trawl nets.

Cetacean bycatch is increasing in intensity and frequency. This is a trend that is likely to continue because of increasing human population growth and demand for marine food sources, as well as industrialization of fisheries which are expanding into new areas. These fisheries come into direct and indirect contact with cetaceans. An example of direct contact is the physical contact of cetaceans with fishing nets. Indirect contact is through marine trophic pathways where fisheries are severely reducing fish stocks that cetaceans rely on for food. In some fisheries, cetaceans are captured as bycatch but then retained because of their value as food or bait. In this fashion, cetaceans can become a target of fisheries.

Bycatch Trends

Generally cetacean bycatch is on the increase. Most of the world's cetacean bycatch occurs in gill-net fisheries. The mean annual bycatch in the U.S. alone from 1990–1999 was 6,215 marine mammals, with dolphins and porpoises being the primary cetaceans caught in gillnets. A study by Read et al. estimated global bycatch through observation of U.S. fisheries and came to the conclusion that an annual estimate of 653,365 marine mammals, comprising 307,753 cetaceans and 345,611 pinnipeds were caught from 1990–1994.

While gillweed nets are a principal concern, other types of nets also pose a problem: trawl nets, purse seines, beach seines, longline gear, and driftnets. Driftnets are known for high rates of bycatch and they affect all cetaceans and other marine species. They are fatal for small toothed whales (*Odontocetes*) and sperm whales, as well as other marine mammals and fish such as sharks, sea birds and sea turtles. Many fisheries routinely use driftnets exceeding the EU size limit of 2.5 km/boat. This illegal drift-netting is a major issue, especially in important feeding and breeding grounds for cetaceans.

However, the tuna industry has achieved successes in reversing cetacean bycatch trends. International recognition of the problem of cetacean bycatch in tuna fishing led to the Agreement on the International Dolphin Conservation Program in 1999 and overall there has been a dramatic reduction in death rates. In particular, dolphin bycatch in tuna fishing in the East Tropical Pacific has dropped from 500,000 per year in 1970 to 100,000 per year in 1990 to 3,000 per year in 1999 to 1,000 per year in 2006.

Cetaceans at Risk

Bycatch is recognized as a primary threat to all cetaceans. The following cetaceans are at high risk for entanglement in gillnets:

Atlantic Humpback Dolphins

The Atlantic humpback dolphin (*Sousa teuszii*) is endemic to West Africa. Several stocks have been identified with numbers ranging from tens to a few hundred. Abundance estimates are lacking. Gaps in the species range and hence distribution is evident. Bycatch is only documented in a few West African countries. Surveys and evaluations need to be conducted to determine the presence/ absence of humpback dolphins in their historical range. Conservation measures need to be implemented to save this species. Because many people live off the sea, it is not feasible to have complete gillnet closures. Some areas may be designated as off-limits to gillnet fisheries. Eco-tourism may be implemented successfully because of high species diversity.

Baleen Whales

Baleen whales, *Mysticeti*, are often taken in gill-nets and in fisheries that use vertical lines to mark traps and pots. Large cetaceans such as humpback and right whales may carry off gear after entanglement. This explains the large scars borne by whales along the U.S. Atlantic coast. Analyses show that 50-70% of Gulf of Maine humpback whales, *Megaptera novaeangliae*, and North Atlantic Right Whale, *Eubalaena glacialis*, have been entangled at least once in their lifetime. The North

Atlantic right whale is one of the most endangered large cetaceans and only 300-350 individuals remain. Minke Whales, *Balaenoptera acutorostrata*, are also at risk.

North Atlantic Right Whale mother and calf.

Burmeister's Porpoises

The Burmeister's porpoise (*Phocoena spinipinnis*) is one of three cetaceans that are most often bycaught in Peru and Chile. Several thousand porpoises are caught each year in Peru alone. Bycatch is a frequent occurrence for this species because of the inability to detect them in the water. Surveys have shown that bycatch remains a concern in that area today and it is unknown whether or not the population is declining. Data, conservation measures and awareness are lacking. These porpoises are cryptic making surveying a challenge . It is also difficult to estimate bycatch because the sale of porpoise meat is no longer available at markets.

Commerson's Dolphins

A Commerson's Dolphin in an aquarium.

The expanding trawl fisheries devastated the Commerson's dolphin (*Cephalorhynchus commersonii*) populations in Patagonia. Trawl fisheries greatly expanded for twenty years until they crashed in 1997. Pelagic squid fisheries took over which use pelagic trawls that are harmful to dusky, short-beaked common dolphins, and Commerson's dolphins. There are approximately 21,000 Commerson's dolphins remaining today. Two stocks have been identified in the population but genetic information and bycatch levels are unknown. With anchovy fisheries expanding, it is imperative to assess the Commerson's dolphin population before these fisheries grow. The seasonal operation of in-shore gillnet fisheries are known to involve bycatch of cetaceans. Presently, there are no known estimates of gillnet bycatch. The bycatch problem in Argentina is political in nature. Improvements in fishing technology, awareness, and a large scale survey of Commerson's dolphin populations and the impact of bycatch is essential.

La Plata Dolphins

The La Plata or Franciscana dolphin (*Pontoporia blainvillei*) is the most threatened small cetacean in the southwest Atlantic Ocean due to bycatch. They are only found in the coastal waters of Argentina, Brazil, and Uruguay. This species has been divided into four ranges (FMU's: Franciscana Management Units) for management and conservation purposes. These populations are genetically different. Mortality rates are 1.6% for FMU 4 and 3.3% for FMU 3 but it is unknown whether these estimates are accurate. Aerial surveys have proven inconclusive so far as to the population numbers of franciscanas. To rectify this situation, more surveys are needed as well as political commitment, awareness campaigns and bycatch mitigation techniques.

Harbour Porpoises

There is substantial incidental catches in fishing operations. Often, the Harbour Porpoise (*Phocoena phocoena*) is killed by incidental by-catch (10, 11, 12). Gillnets pose a serious threat to the harbour porpoise as they are extremely susceptible to entanglement. A study by Caswell et al. in the western North Atlantic combined the mean annual rate of increase of the harbour porpoise with the uncertainty of incidental mortality and population size. It was found that the incidental mortality exceeds critical values and therefore by-catch is a significant threat to the harbour porpoise. Harbour porpoises become entangled in nets due to their inability to detect the nets before collision. In 2001, 80 harbour porpoises were killed in salmon gillnet fisheries in British Columbia, Canada.

Hector's and Maui's Dolphins

Hector's dolphins have a unique rounded dorsal fin.

In New Zealand, these dolphins have a high rate of entanglement. Hector's dolphin (*Cephalorhynchus hectori*) is endemic to the coastal waters New Zealand and there are about 7,400 in abundance. A small population of Hector's dolphins is isolated on the west coast of the island and have been declared a subspecies called Maui's Dolphin. Maui's dolphins (*Cephalofhynchus hectori maui*) are often caught in set nets and pair trawlers resulting in less than 100 left in the wild. For protection, a section of the dolphin's range on the west coast has been closed to gillnet fisheries.

Indo-Pacific Humpback and Bottlenose Dolphins

Drift and bottom-set gillnets are the biggest conservation threat to these dolphins in the Indian Ocean. There have only been assessments in some areas, such as Zanzibar. Hunting, until 1996, reduced the population and contributed to its decline. Now hunting has been replaced with eco-tourism. It was estimated in 2001 that there are 161 bottlenose dolphins (*Tursiops aduncus*) and 71 Indo-Pacific Humpback Dolphin (*Sousa chinensis*) that are left based on photo-identification mark-recapture techniques. A study on bycatch revealed over 160 incidences of bycatch since 2000. Approximately 30% of bycatch is in drift and bottom-set gillnets. Mortality is about 8% and 5.6% for bottlenose and humpback dolphins respectively . The mitigation of bycatch is imperative for these species and eco-tourism.

Irrawaddy Dolphins

Based on a survey in 2001, fewer than 70 Irrawaddy dolphins (*Orcaella brevirostris*) left in the upper region of the Malampaya Sound in the Philippines and 69 individuals in the Mekong River. They have been severely impacted by lift nets, and crab gear and they are critically endangered. It is estimated that mortality from bycatch may be greater than 4.5% in Malampaya Sound and 5.8% in the Mekong River. The population is declining dramatically. Current bycatch levels are unsustainable and bycatch reduction measures as well as long-term systematic monitoring are urgently required. The elimination of gillnets from areas of high use is needed and economic incentives need to be provided to the local people.

Spinner and Fraser's Dolphins

Spinner dolphins.

In the Philippines, tuna driftnet fisheries have a substantial impact on the populations. One tuna fishery alone kills 400 Spinner Dolphin (*Stenella longirostris*) and Fraser's dolphins (*Lagenodelphis hosei*) each year. Round-haul nets are an even greater concern with a bycatch of up to 3000 dolphins per year. Dolphins that are bycaught often end up as shark bait for longline fisheries. There is not enough data to conclude total bycatch for the Philippines. Initial assessment indicates that bycatch is not sustainable. Monitoring of dolphin populations and fisheries is urgently needed.

Yangtze River Dolphins and Finless Porpoises

Illustration of a Baiji dolphin.

The Yangtze River or Baiji dolphin (*Lipotes vexillifer*) is the most endangered cetacean and is only found in the Yangtze River, China. A survey conducted in 1997 found only thirteen dolphins. The Yangtze River finless porpoise (*Neophocaena phocaenoides asiaeorientalis*) also lives in the Yangtze River. Abundance has declined and there are fewer than 2000 dolphins left. This may be due, in part, to the construction of the Three Gorges Dam which covers a significant amount of the dolphin's habitat. Both species are often subject to entanglement in gillnets.

Vaquita

The vaquita (*Phocoena sinus*) is highly endangered and is endemic to the upper Gulf of California, Mexico. They are killed in both gillnets and trawl nets from commercial and artisanal fishing. There are presently less than 100 vaquitas left in the Gulf of California.

Mitigating Bycatch

Acoustic Deterrent Devices

The use of acoustic alarms to mitigate by-catch and also to protect aquaculture sites has been proposed but has advantages and risks associated with the alarms. Acoustic deterrent devices, or pingers, have reduced the number of cetaceans caught in gill nets. Harbour porpoises have been effectively excluded from bottom-set gill nets during many experiments for instance in the Gulf of Maine, along the Olympic Peninsula, in the Bay of Fundy, and in the North Sea. All of these studies show up to a 90% decrease in harbour porpoise bycatch. Pingers work because they produce a sound that is aversive (20; 15). There has been a recent re-evaluation of the potential of pingers and their use in other fisheries due to their growing success. An experiment on the California drift gill net fishery demonstrated how acoustic pingers reduce marine mammal bycatch. It was shown that bycatch was significantly reduced for common dolphins and sea lions. Bycatch rates were also lower for other cetacean species like the Northern right whale dolphin, Pacific white-sided dolphin, Risso's dolphin and Dall's Porpoise. It is agreed upon that the more pingers on a net, the less bycatch. There was a 12-fold decrease in common dolphin entanglement using a net with 40 pingers. However, the widespread use of pingers along coastlines effectively excludes cetaceans such as porpoises from prime habitat and resources. Cetaceans which are extremely sensitive to noise are effectively being driven from their preferred coastal habitats by the use of acoustic devices. In poorer quality habitat, harbour porpoises are subjected to increased competition for resources. This situation is recognized as range contraction which can be a result of climate change, anthropogenic activity, or population decline. Large scale range contractions are considered indicative of impending extinction. A similar form of deterrent is noise pollution originating from vessel traffic.

Barium Sulfate

A promising gillnet that is effective in reducing bycatch for harbor porpoises contains barium sulfate. These nets are detected at a greater distance than conventional nets because the barium sulfate reflects the echolocation signal, and also renders the nets more visible. Barium sulfate makes the nets stiffer if it is added at high concentration. All three factors: echo reflectivity, stiffness, and visibility may be important in reducing bycatch. Fish takes in the Bay of Fundy were normal, except for haddock takes, which were down by 3-5%. The advantage of this approach is that it is passive and thus does not require batteries, and there is no "dinner bell" effect. The potential advantage of these nets is greatest in the artisanal fishery. NOAA would like further testing to verify the effectiveness of the nets.

Fishing Regulations and Management

Management and regulation are lacking in many fisheries today. Management measures are urgently needed to monitor fisheries (and illegal fisheries) to protect cetaceans. Efforts to document bycatch should focus on gill-net fisheries because cetaceans are more likely to be caught in gill-nets. Conservation efforts should be directed to areas where marine mammal bycatch is high but where no infrastructure exists to assess the impact. There is a lack of reporting on a global scale of cetacean bycatch.

In the U.S. the Marine Mammal Protection Act prohibits the use and sale of marine mammals captured by fisheries. Similar legislation prohibits the use and sale of marine mammals in other countries. A marine mammal mortality monitoring program for commercial fisheries occurs in the U.S. where "Take Reduction Teams" observe the extent of bycatch and then formulate strategies to reduce bycatch and Take Reduction Plans are put into place.

Temporary Closure

Temporary closure of fisheries during the short period of the year when cetaceans are migrating through the area would decrease bycatch significantly.

In the U.S.

Some programs like Earth Island Institute's Dolphin Safe Label certification claim to require certification from onboard observers. However, the only fishery in the world where independent scientific observers certify whether or not a dolphin has been harmed is the Eastern Tropical Pacific, home to the AIDCP Treaty program. For all other tuna fisheries of the world, the efficacy of onboard observer certification has come under increasing scrutiny as such programs have proven indefensible or unmanageable:

In an interview with Radio Australia last year, Mark Palmer of EII confirmed that it is mostly the case that EII monitors do not go on board of the vessels, and their organization does not have the kind of resources to put observers on the "many thousands" of ships that are out there catching tuna.

Additionally, environmental groups have criticized Earth Island Institute's support of U.S. policies that do not require independent, on-board observation and instead only rely on self-certification by fishing captains, and that even where they may at some point in the future require independent observers, the lack of uniformity in tracing and verifying certifications in different countries means non-certified products can become certified if they are simply taken to the right port.

Other ways of Mitigating Bycatch

- Implement gear technology (changes in fishing gear and practices) documented to mitigate cetacean bycatch.

- Buy tuna and other seafood that has a dolphin safe label.

- Buy Sustainable seafood. To find out which seafood is produced sustainably (i.e. using cetacean friendly gear), refer to World Wildlife Fund Global to access worldwide sustainable seafood guides.

- Support sustainable seafood companies and restaurants.

- Raise international awareness to assess, monitor and mitigate bycatch problems.

- Create legislation on responsible fishing practices.

- Develop and promote industry adoption of "Best Practice Guidelines" for fishing operations.

Reclamations

Land reclamations can destroy nursery grounds of juvenile fish and the habitats of shellfish. Many coastal cities e.g. Wellington, Auckland and Lyttelton, have reclaimed shallow marine areas for extra land. Reclamation is still continuing around New Zealand, but now environmental Impacts Reports are required as part of the consent process. Regional coastal plans are also required to be prepared by regional councils, setting out what activities and effects are allowed in the coastal marine are. The impact or reclamation on fisheries resources is more likely to be taken into account.

Mineral Exploitation

Oil production, with its associated pollution risks, and mining of the seabed and beach sands can disrupt marine habitats, Mining companies are now much more aware of environmental impacts than in the past.

Pollution

Dealing with waste products is a worldwide problem for both developed and developing countries, Many cities and large industries have often chosen the " out of sight, out of mind" option, discharging sewage and waste products into the sea, Some of the waste products are toxic. In parts of the North Sea, some fish are now so contaminated they're unfit to eat. Some fish are being born deformed.

Rubbish

Rubbish, such as plastic and bits of fishing nets dumped at sea or on the foreshore is a menace to fish, marine mammals and birds. Fishers are often blamed for this, and while some probably are still irresponsible.

Enhancement (Reseeding)

Enhancement is another more positive human impact. It involves releasing hatchery-reared young into the wild or providing additional protection to naturally spawned juveniles. This is not done on a wide scale, because of cost, the exceptions in New Zealand being scallops and salmon. Research is being done on snapper and rock lobster enhancement.

Global Warming

A changing climate will affect the marine environment, altering sea levels, temperature and salinity, current direction and strength, nutrient level and the nature and distribution of the boundaries between water masses. These changing conditions will change the distribution, reproduction and growth of many fish species.

Ozone Depletion and Ultraviolet Impacts

UV-b is the most harmful component of ultraviolet radiation. A higher level of UV-b radiation is reaching the earth because of the reduction in the ozone layer. Scientists have not yet been able to predict reliably the effects of this on marine life, but there is increasing worldwide concern about the impact on plankton and marine ecosystems.

Comparing Effectiveness of Experimental and Implemented Bycatch reduction Measures: the Ideal and the Real.

Fishers, scientists, and resource managers have made substantial progress in reducing bycatch of sea turtles, seabirds, and marine mammals through physical modifications to fishing gear. Many bycatch-avoidance measures have been developed and tested successfully in controlled experiments, which have led to regulated implementation of modified or new fishing gear. Nevertheless, successful bycatch experiments may not translate to effective mitigation in commercial fisheries because experimental conditions are relaxed in commercial fishing operations. Such a difference between experimental results and real-world results with fishing fleets may have serious consequences for management and conservation of protected species taken as bycatch. We evaluated reimplementation experimental measures and post implementation efficacy from primary and gray literature for three case studies: acoustic pingers that warn marine mammals of the presence of gill nets, turtle excluder devices that reduce bycatch of turtles in trawls, and various measures to reduce seabird bycatch in longlines.

NOAA's National Marine Fisheries Service (NMFS) Office of Protected Resources cooperates with partners to conserve and recover protected marine species by minimizing human impacts. Below are some examples.

Fisheries Interactions (Bycatch):

The Office of Protected Resources' Fisheries Interactions program works to implement section

118 of the Marine Mammal Protection Act (MMPA) and regulations governing the incidental capture of marine mammals in commercial fisheries. The Office works with NMFS' regional offices through take reduction planning to reduce marine mammal bycatch in commercial fisheries. Additionally, the Office works to reduce bycatch of marine turtles by implementing management measures such as time/area closures, modifications to fishing gear and practices, and safe sea turtle handling practices.

Ocean Sound/Acoustics:

Intense underwater sound can harass or harm marine mammals. Human sources of sound include military activities, vessel operations, petrochemical and geophysical exploration, marine construction, and research activities. NMFS experts review proposed underwater activities and develop solutions to minimize the potential impacts to marine mammals.

Ship Strikes:

Many species of marine mammals are injured by ship strikes. The problem is greatest for the critically endangered Northern right whale, one of the most commonly struck species. The Office of Protected Resources, in cooperation with the U.S. Coast Guard, works to minimize ship strikes by relaying information about recent whale locations to mariners. In addition, large ships are required to report when they enter areas with a high risk of a ship strike.

Viewing Wildlife:

Viewing marine animals can be educational and enriching when conducted responsibly. However, without certain precautions, these activities can put both the animals and the viewers at risk. The Office of Protected Resources promotes responsible wildlife viewing through posters, pamphlets, and workshops.

Safely Deterring Marine Mammals:

Human-marine mammal interactions are usually considered from the perspective of the impact of human activity on marine mammal populations or individuals. However, marine mammals may have an effect on human activities or property. As a result of such conflict, Congress included a provision for deterring marine mammals in section 101 of the MMPA to allow certain people to use safe, non-lethal methods to deter marine mammals to protect private or public property. (NMFS, in collaboration with other Federal and state officials, has prepared information for deterring marine mammals, which is available on page at NMFS' Northwest Regional Office and NMFS' Southwest Regional Office.)

Human Impacts on Biodiversity of Aquatic Ecosystem

Aquatic ecosystems contribute to a large proportion of the planet's biotic productivity as about 30% of the world's primary productivity comes from plants living in the, ocean. These ecosystems also include wetlands located at lakeshores, riverbanks, the ocean shoreline, and any habitat where the soil or vegetation is submerged for some duration. When compared to terrestrial communities, aquatic communities are limited abiotically in several different ways.

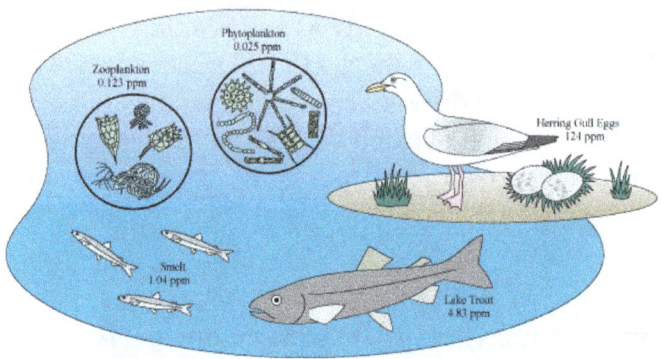

- Organisms in aquatic systems survive partial to total submergence. Water submergence has an effect on the availability of atmospheric oxygen, which is required for respiration, and solar radiation, which is needed in photosynthesis.

- Some organisms in aquatic systems have to deal with dissolved salts in their immediate environment. This condition has caused these forms of life to develop physiological adaptations to deal with this problem.

- Aquatic ecosystems are nutritionally limited by phosphorus and iron, rather than nitrogen and,

- These are generally cooler than terrestrial systems, which limit metabolic activity.

Global Scenario

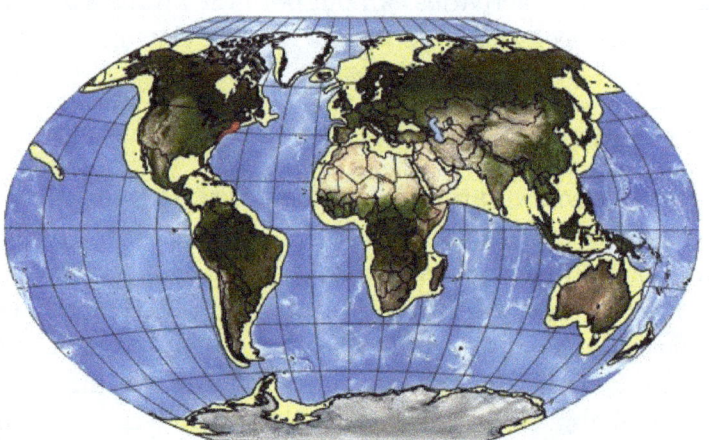

The earth, two-thirds of which is covered by water, looks like a blue planet-the planet of water-from space (Clarke, 1994). The world's lakes and rivers are probably the planet's most important freshwater resources. But the amount of fresh water constitutes only 2.53% of the earth's water. On the earth's surface, fresh water is the habitat of a large number of species. These aquatic organisms and the ecosystem in which they live represent a substantial sector of the earth's biological diversity. The association of man and aquatic ecosystem is ancient. It is not surprising that the first sign of civilization is traced to wetland areas. The flood plains of the Indus, the Nile delta, and the Fertile Crescent of the Tigris and Euphrates rivers provided man with all his basic necessities. Water may be required for various purposes like drinking and personal hygiene, fisheries, agriculture, navigation, industrial production, hydropower generation, and recreational activities. The wide variety of wetlands, like marshes, swamps, bogs, peat land, open water bodies like lakes and rivers, mangroves, tidal marshes, and so forth, can be profitably used by humans for various needs and for environmental amelioration. Ever-increasing population and the consequent urbanization and industrialization have mounted serious environmental pressures on these ecosystems and have affected them to such an extent that their benefits have declined significantly.

It is interesting to know that there are nearly 14×10^8 cubic km of water on the planet, of which more than 97.5% is in the oceans, which covers 71% of the earth's surface. Wetlands are estimated to occupy nearly 6.4% of the earth's surface. Of those wetlands, nearly 30% is made up of bogs, 26% fens, 20% swamps, and 15% flood plains. Of the earth's fresh water, 69.6% is locked up in the continental ice, 30.1% in underground aquifers, and 0.26% in rivers and lakes. In particular, lakes are found to occupy less than 0.007% of world's fresh water (Clarke, 1994).

References

- Barron MG, Carls MG, Heintz R, Rice SD. 2003. Evaluation of fish early-life stage toxicity models of chronic embryonic exposures to complex polycyclic aromatic hydrocarbon mixtures. Oxford Journals. 78(1):60-67

- Rahim, Mia; Islam, Tarikul; Kuruppu, Sanjaya (July 2016). "Regulating global shipping corporations' accountability for reducing greenhouse gas emissions in the seas". Marine Policy. 69: 159–170. doi:10.1016/j.marpol.2016.04.018

- Worm, Boris; et al. (2006-11-03). "Impacts of Biodiversity Loss on Ocean Ecosystem Services". Science. 314 (5800): 787–790. PMID 17082450. doi:10.1126/science.1132294. Retrieved 2006-11-04

- Hussein, Abdel, and Mona Mansour. 2015. A review on polcyclic aromatic hydrocarbons: Source, environmental impact, effect on human health and remediation. Egyptian Journal of Petroleum 25: 107-123

- Bolster, W. Jeffery (2012). The Mortal Sea: Fishing the Atlantic in the Age of Sail. Belknap Press. ISBN 978-0-674-04765-5

- Vestergaard, N.; Squires, D.; Kirkley, J.E. (2003). "Measuring Capacity and Capacity Utilization in Fisheries. The Case of the Danish Gillnet Fleet". Fisheries Research. 60 (2–3): 357–68. doi:10.1016/S0165-7836(02)00141-8

- Human Noise Pollution in Ocean Can Lead Fish Away from Good Habitats and Off to Their Death, University of Bristol, 13 August 2010, retrieved 2011-03-06

- Gordon, H. S. (1953). "An Economic Approach to the optimum utilization of Fishery Resources". Journal of the Fisheries Research Board of Canada. 10 (7): 442–57. doi:10.1139/f53-026

- Hall, M; Alverson, DL; Metuzals, KI (2000). "By-catch: problems and solutions". Marine Pollution Bulletin. 41 (1–6): 204–219. doi:10.1016/S0025-326X(00)00111-9

- Scales, Helen (29 March 2007). "Shark Declines Threaten Shellfish Stocks, Study Says". National Geographic News. Retrieved 2012-05-01

- Clucas, I.; Teutscher, F., eds. (1999). FAO/DFID Expert Consultation on Bycatch Utilization in Tropical Fisheries. Beijing (China), 21–28 Sep 1998. University of Greenwich, NRI. p. 333. ISBN 0-85954-504-0

- Murray, KT, Read, AJ, and AR Solow. 2000. The use of time/area closures to reduce bycatches of harbour porpoises: lessons from the Gulf of Maine sink gillnet fishery. Journal of Cetacean Research and Management. 2(2): 135-141

- Brothers NP (1991). "Albatross mortality and associated bait loss in the Japanese longline fishery in the southern ocean". Biological Conservation. 55 (3): 255–268. doi:10.1016/0006-3207(91)90031-4

- "The Top 10 Everything of 2009: Top 10 Scientific Discoveries: 5. Breeding Tuna on Land". Time. 8 December 2009. Retrieved 2012-05-01

- Demaster, DJ, Fowler, CW, Perry, SL, and ME Richlen (2001). Predation and competition: the impact of fisheries on marine mammal populations over the next one hundred years. Journal of Mammalogy. 82: 641-651

- Nada, M., & Casale P. (2011). Sea turtle bycatch and consumption in Egypt threatens Mediterranean turtle populations. Fauna & Flora International, Oryx, 45(1), 148. doi : 10.1017/S0030605310001286

Marine Conservation and Management

Important areas of marine conservation include habitat alteration and loss, species introduction of aquatic organisms, organisms facing extinction, etc. Examples of marine ecosystem management in different regions have been discussed. The aspects elucidated in this chapter are of vital importance, and provide a better understanding of marine conservation and management.

Marine Conservation

Aquatic biodiversity includes variety of life and ecosystems of freshwater, brackish water and marine environment. The human societies had long been depending upon aquatic biodiversity for food, medicine and other uses including commercial and industrial nature. The economic value of aquatic biodiversity is immeasurable and immense.

In recent times, the factors like over-exploitation, pollution, habitat alteration and destruction, introduction of alien species etc., are overwhelmingly causing impacts and threats to aquatic biodiversity.

There is necessity to put in place appropriate conservation strategies and actions to safeguard the aquatic biodiversity for the benefit of the present, as well as, future generations. The conservation oriented scientific pursuit and technology backed interventions would only address the pressing problems in protecting aquatic biodiversity and its sustainable use with the understanding of interdependence of organisms and ecosystems and human needs in the present day context.

Ecologically effective ecosystem management will require the development of a robust logic, rationale, and framework for addressing the inherent limitations of scientific understanding. It must incorporate a strategy for avoiding irreversible or large-scale environmental mistakes that arise from social and political forces that tend to promote fragmented, uncritical, short-sighted, inflexible, and overly optimistic assessments of resource status, management capabilities, and the consequences of decisions and policies.

Aquatic resources are vulnerable to the effects of human activities catchment-wide, and many of the landscape changes humans routinely induce cause irreversible damage (e.g., some species introductions, extinctions of ecotypes and species) or give rise to cumulative, long-term, large- scale biological and cultural consequences (e.g., accelerated erosion and sedimentation, deforestation, toxic contamination of sediments). In aquatic ecosystems, biotic impoverishment and environmental disruption caused by past management and natural events profoundly constrain the ability of future management to maintain biodiversity and restore historical ecosystem functions and values. To provide for rational, adaptive progress in ecosystem management and to reduce the risk of irreversible and unanticipated consequences, managers and scientists must identify catchments and aquatic networks where ecological integrity has been least damaged by prior management, and jointly develop means to ensure their protection as reservoirs of natural biodiversity, keystones for regional restoration, management models, monitoring benchmarks, and resources for ecological research.

Marine conservation is the protection and preservation of ecosystems in oceans and seas. Marine conservation focuses on limiting human-caused damage to marine ecosystems, restoring damaged marine ecosystems, and preserving vulnerable species of the marine life.

Overview

Marine conservation is a response to biological issues such as extinction and marine habitats change. Marine conservation is the study of conserving physical and biological marine resources and ecosystem functions. This is a relatively new discipline. Marine conservationists rely on a combination of scientific principles derived from marine biology, oceanography, and fisheries science, as well as on human factors such as demand for marine resources and marine law, economics and policy in order to determine how to best protect and conserve marine species and ecosystems. Marine conservation can be seen as sub-discipline of conservation biology.

Conservation of Aquatic Biodiversity

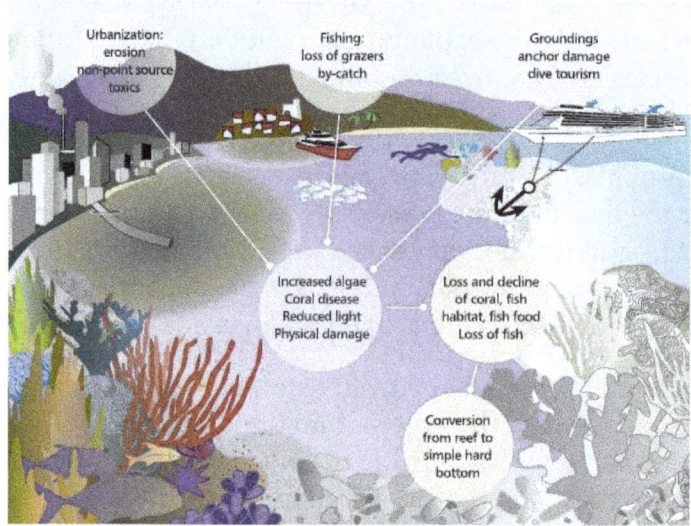

The streams of the southern Appalachians drain ancient landscapes (>250million years old), and contain a rich and distinctive fauna with many endemics. The areas long evolutionary history has not been interrupted by sea level rise or Pleistocene glaciations.

The level of aquatic biodiversity is widely recognized as the highest of any temperate region in the world, rivaling tropical systems. The diversity of aquatic invertebrate species appears to be greater than that of any other region in North America, with up to 50 percent of some taxonomic groups still undescribed. Out of 297 mussel species occurring in the United States, 269 are found in the Southeast. An estimated 350 species of fish occur, some 18 percent of which are imperiled.

The diverse fauna and its setting in a rapidly changing landscape present substantial challenges for aquatic resource managers. Conservation of individual species, aquatic communities, and flowing water habitats will be potentially difficult. Over the past century a large body of knowledge has accumulated on the zoogeography, distribution, and biology of the southeastern fish fauna and to a lesser extent other aquatic organisms. This diversity is threatened. A number of imperiled species from all faunal groups are threatened or endangered. The primary threats to the biological integrity of the region are habitat alteration and loss and introduced species.

Subsection Found in Conservation of Aquatic Biodiversity

- Imperiled Aquatic Species

- Habitat Alteration and Loss

- Introduced Species in Aquatic Systems

Imperiled species are those believed to be at some risk of extirpation or extinction; here, threatened, endangered, and special concern species are collectively referred to as imperiled species. Threatened and endangered (T&E) species have officially been listed by the U.S. Fish and Wildlife Service (FWS) under the Endangered Species Act of 1973. Special concern (SC) species may be limited in distribution and abundance, but the legal listing process has not been completed; these species are recognized by NatureServe as having globally limited distributions (G1, G2, G3). Most of the imperiled species in the southern Appalachian region are in decline due to habitat alteration and loss.

Imperiled fishes, molluscs, reptiles, amphibians, and invertebrates in the southern Appalachian region include species that are federally protected and those that are globally rare.

Introduced Species in Aquatic Systems

Nonnative or exotic species are those that have been introduced by humans, intentionally or unintentionally, to ecosystems in which they did not evolve. Introduced species are a global problem. The establishment of exotic species in United States freshwater is on the rise. In 1920, six exotic fishes were established; just three more had been added by 1945. By 1980, an estimated 35 species were established and approximately 50 more had been observed.

Introduced species are special problems for several reasons. Once alien species are established in a new environment, they often are capable of reproducing and spreading far beyond the point of entry. Unlike chemical pollutants that can be eliminated at the source, or habitats that might potentially be restored, species introductions are usually impossible to reverse. Effects of exotic species on native stream organisms can include predation, competition, habitat alterations, hybridization, and introduction of disease or parasites.

Other causes of species invasions have included:

- Intentional introductions by new settlers nostalgic for familiar species.

- Intentional introductions by government agencies to supply sport or food organisms to the public. (For example, stocking of nonnative trout continues today).

- Intentional introductions by individual citizens to augment local fauna.

- Intentional and unintentional releases by aquarists and pond owners.

- Unintentional (or unthinking) releases from bait buckets.

The reduction in the distribution of native brook trout is a well-known example of the effects of introduced species. This species was once abundant in cold mountain streams above about 2,000 foot elevation. Introduced rainbow trout began encroaching on brook trout streams around 1900 and brown trout by mid-20th century. Rainbow trout dominated Great Smoky Mountains National Park by 1942. Total miles of brook trout-only streams decreased by about 45 percent from then until the 1970s. Brook trout remain restricted to headwaters in North Carolina, Tennessee, and Georgia, but are more widely distributed further north. The mechanism for the replacement has not been identified, but interference competition has been implicated.

Habitat Alteration and Loss

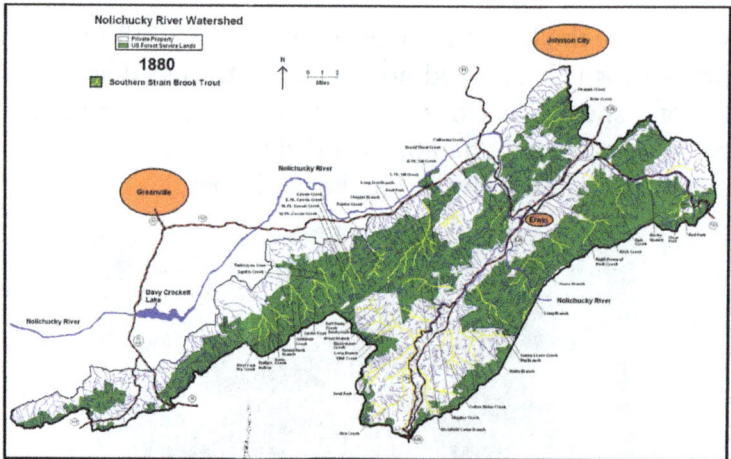

Because southern Appalachian rivers were not exposed to Pleistocene glaciation, native aquatic organisms have had a long evolutionary history, and possess adaptations to the environmental conditions that prevail in mountain rivers and streams. These conditions reflect the natural forest cover of landscapes in the region, and include:

- Extensive physical structure as provided by rocky stream beds and large woody debris.

- Clear, pure water with low concentrations of dissolved nutrients and suspended material.

- Cool to cold water depending on elevation.

Most impacts to aquatic ecosystems indirectly affect organisms by altering their habitat. The most harmful impacts can render the habitat unlivable for sensitive species. Habitat alteration may include any combination of the following:

- Modification of natural flow regimes by dams, diversions, and watershed land uses.

- Changes in water quality due to point and nonpoint source pollution.

- Woody debris inputs to streams are diminished or lost as a result of reduced forest cover along waterways.

The combined, possibly synergistic, effect of these changes in the physical and chemical environment amounts to a deterioration in the quality of habitat for native taxa. Species tolerant of environmental change dominate biological communities, while native species decline. Endemic species, whose distribution is often limited to a single watershed, appear to be particularly sensitive to environmental changes, disappearing where habitats have been altered extensively. Endemics are often disproportionately listed as threatened or endangered. As habitat continues to be degraded under the socioeconomic pressures of development, the threats to the well-being of native organisms and ecosystems mount. Loss of endemic species and concomitant invasion of tolerant, generalist species from outside the region erode the unique faunal diversity of the region. This trend toward homogenization points to a simplified, ecologically bleak future if actions are not taken to reverse the process.

Proper stream habitat management and best management practices in streamside zones can help protect and restore aquatic ecosystems.

Ecosystem Approach

Aquatic biodiversity plays a vital role in rural livelihoods. However, it is being threatened by factors within the fisheries sector, such as overfishing, destructive fishing practices and introduction of alien species, as well as by external factors such as habitat loss and degradation mainly caused by land-based activities. Thus, the FAO Aquaculture Management and Conservation Service embarked on a programme aimed at constructing an inventory and valuation of inland aquatic biodiversity that is used by rural communities in natural and modified ecosystems with special emphasis on traditional knowledge, sustainable use, enhancement and gender issues.

The conservation and sustainable use of fish stocks need to be promoted urgently by linking ecosystem considerations into capture fisheries management practices and procedures. A set of guidelines on ecosystem approaches to fisheries management has been developed by FAO.

FAO Code of Conduct for Responsible Fisheries

The FAO Fisheries and Aquaculture Department is carrying out a variety of activities in relation to aquatic biodiversity that are considered essential for sustainable fisheries and aquaculture. The 1982 United Nations Convention on the Law of the Sea (UNCLOS 1982) and the FAO Code of Conduct for Responsible Fisheries (CCRF 1995) provide the umbrella for FAO's work in fisheries.

Conservation of Aquatic Biodiversity in Vietnam

The ecosystem approach is a conceptual framework for resolving ecosystem issues. The idea is to protect and manage the environment through the use of scientific reasoning. Another point of the ecosystem approach is preserving the Earth and its inhabitants from potential harm or permanent damage to the planet itself. With the preservation and management of the planet through an ecosystem approach, the future monetary and planetary gain are the by-product of sustaining and/or increasing the capacity of that particular environment. This is possible as the ecosystem approach incorporates humans, the economy, and ecology to the solution of any given problem. The initial idea for an ecosystem approach would come to light during the second meeting (November 1995) at the Conference of the Parties (COP) it was the central topic in implementation and framework for the Convention on Biological Diversity (CBD), it would further elaborate on the ecosystem approach as using varies methodologies for solving complex issues.

Throughout, the use and incorporation of ecosystem approaches, two similar terms have been created in that time: ecosystem-based management and ecosystem management. The Convention on Biological Diversity has seen ecosystem-based management as a supporting topic/concept for the ecosystem approach. Similarly, ecosystem management has a minor difference with the two terms. Conceptual the differences between the three terms come from a framework structure and the different methods used in solving complex issues. The key component and definition between the three terms refer to the concept of conservation and protection of the ecosystem.

The use of the ecosystem approach has been incorporated with managing water, land, and living organisms ecosystems and advocating the nourishment and sustainment of those ecological space. Since the ecosystem approach is a conceptual model for solving problems, the key idea could combat various problems.

History

On December 29, 1993, the Convention on Biological Diversity (CBD) was signed and applied as a multilateral treaty. With the purpose of achieving:

1.   Biodiversity

2. Sustainability of species diversity

3. Endorse genetic diversity (e.g. to maintain and endorses livestock, crops, and wildlife)

Two years after the CBD was signed, during the second meeting of the Conference of the Parties (November 1995) the representatives of the signed treaty would agree upon employing a strategy to combat the intricate and actively changing ecosystem. The ecosystem approach would represent as the equalizer for obtaining knowledge and creating countermeasures in preventing the endangerment of any ecological environment.

With the acknowledgment of the ecosystem approach, during the fifth meeting of the Conference of the Parties, a group consensus agreed on a concrete definition and elaboration for the ecosystem approach would be needed, and the Parties would request Subsidiary Body on Scientific Technical and Technological Advice (SBSTTA) to create a guideline with 12 principles and a description of the ecosystem approach. The final results are given at COP 5 Decision V/6 summary.

During the seventh meeting of the Conference of the Parties, further iteration on the ecosystem approach would be seen as a priority, during the meeting the parties would agree new implementation and strategical development could be incorporated with the ecosystem approach into the CBD. Furthermore, creating a new relationship with sustaining forest organization and the ecosystem approach was talked about.

Ecosystem Approach and Management

With the development and use of the ecosystem approach, different variation to that form have been created and used. The two being ecosystem management and ecosystem-based management, the framework of the three methods are still the same (the conservation and protection of the ecosystem). The distinguishing part beings with how to initiate the approach of solving the problem. Ecosystem-based management (EBM) is used for projects that incorporate interaction of different levels: organisms, the ecosystem, and the human component; however, its varies from the other methods as the scale of the problem is larger and intricate. The objectives should be straightforward and condensed with important systematic information. Also, EBM considers social and cultural aspects into the solution not just only scientific reasoning. With ecosystem management, the process is similar to EBM; however, factors such as socioeconomics and politics can impact the decision and solution. A cultural aspect is also considered when creating a solution.

Ecosystem Approach to Fisheries

The ecosystem approach is currently being used in the fields of environmental and ocean management (i.e. the ecosystem approach does not stop there; it is being used in various fields and sub-fields as well). The goal is to address the current problems facing those fields through the use of conceptual thinking and approach that would determine a viable and sustaining solution. One example, in particular, would be pertaining to Fishery (a commercial industry in capturing and selling of fishes). In recent years, inland fisheries have quintupled from 2 million metric tons to 11 million metric tons in the span of 60 years from 1950 to 2010. Through the use of the ecosystem approach more specifically the ecosystem approach to fisheries (EAF), sometimes referred to as Ecosystem-based fisheries. EAF is seen as a framework to creating local strategies for each specific fishery ecosystem and implementing the new strategies gradually with the already existing rules and regulation. With the use of EAF if successful fishery industries could generate substantial income; as well as, improve the fragile ecosystem of aquarium species.

Sustainable Fishery

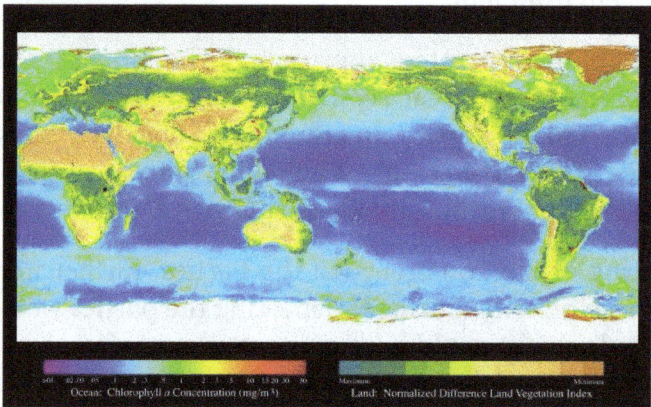

SeaWiFS map showing the levels of primary production in the world's oceans

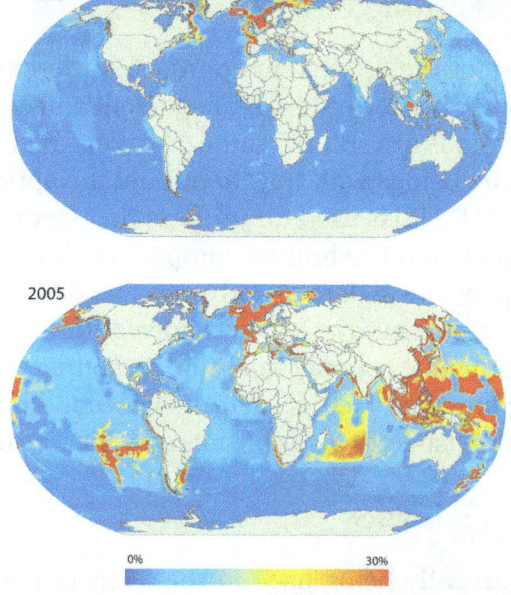

Primary production required (PPR) to sustain global marine fisheries landings expressed as percentage of local primary production (PP).Estimates of PPR, PP and PPR/PP computed per 0.5° latitude/longitude ocean cells. PPR estimates based on the catch database and PP estimates derived from SeaWiFS's global ocean colour satellite data. The maps represent total annual landings for 1950 (top) and 2005 (bottom). Note that PP estimates are static and derived from the synoptic observation for 1998.

A conventional idea of a sustainable fishery is that it is one that is harvested at a sustainable rate, where the fish population does not decline over time because of fishing practices. Sustainability in fisheries combines theoretical disciplines, such as the population dynamics of fisheries, with practical strategies, such as avoiding overfishing through techniques such as individual fishing quotas, curtailing destructive and illegal fishing practices by lobbying for appropriate law and policy, setting up protected areas, restoring collapsed fisheries, incorporating all externalities involved in harvesting marine ecosystems into fishery economics, educating stakeholders and the wider public, and developing independent certification programs.

Some primary concerns around sustainability are that heavy fishing pressures, such as overexploitation and growth or recruitment overfishing, will result in the loss of significant potential yield; that stock structure will erode to the point where it loses diversity and resilience to environmental fluctuations; that ecosystems and their economic infrastructures will cycle between collapse and recovery; with each cycle less productive than its predecessor; and that changes will occur in the trophic balance (fishing down marine food webs).

Overview

"Sustainable management of fisheries cannot be achieved without an acceptance that the long-term goals of fisheries management are the same as those of environmental conservation".

Daniel Pauly and Dave Preikshot,

Global wild fisheries are believed to have peaked and begun a decline, with valuable habitats, such as estuaries and coral reefs, in critical condition. Current aquaculture or farming of piscivorous fish, such as salmon, does not solve the problem because farmed piscivores are fed products from wild fish, such as forage fish. Salmon farming also has major negative impacts on wild salmon. Fish that occupy the higher trophic levels are less efficient sources of food energy. Fishery ecosystems are an important subset of the wider marine environment.

History

In the end, we will conserve only what we love; we will love only what we understand; and we will understand only what we are taught

Senegalese conservationist Baba Dioum,

In his 1883 inaugural address to the International Fisheries Exhibition in London, Thomas Huxley asserted that overfishing or "permanent exhaustion" was scientifically impossible, and stated that probably "all the great sea fisheries are inexhaustible". In reality, by 1883 marine fisheries were already collapsing. The United States Fish Commission was established 12 years earlier for the purpose of finding why fisheries in New England were declining. At the time of Huxley's address, the Atlantic halibut fishery had already collapsed (and has never recovered).

Traditional Management of Fisheries

Traditionally, fisheries management and the science underpinning it was distorted by its "narrow focus on target populations and the corresponding failure to account for ecosystem effects leading to declines of species abundance and diversity" and by perceiving the fishing industry as "the sole legitimate user, in effect the owner, of marine living resources." Historically, stock assessment scientists usually worked in government laboratories and considered their work to be providing services to the fishing industry. These scientists dismissed conservation issues and distanced themselves from the scientists and the science that raised the issues. This happened even as commercial fish stocks deteriorated, and even though many governments were signatories to binding conservation agreements.

Defining Sustainability

The notion of sustainable development is sometimes regarded as an unattainable, even illogical notion because development inevitably depletes and degrades the environment.

Ray Hilborn, of the University of Washington, distinguishes three ways of defining a sustainable fishery:

- *Long term constant yield* is the idea that undisturbed nature establishes a steady state that changes little over time. Properly done, fishing at up to maximum sustainable yield allows nature to adjust to a new steady state, without compromising future harvests. However, this view is naive, because constancy is not an attribute of marine ecosystems, which dooms this approach. Stock abundance fluctuates naturally, changing the potential yield over short and long term periods.

- *Preserving intergenerational equity* acknowledges natural fluctuations and regards as unsustainable only practices which damage the genetic structure destroy habitat, or deplete stock levels to the point where rebuilding requires more than a single generation. Providing rebuilding takes only one generation, overfishing may be economically foolish, but it is not unsustainable. This definition is widely accepted.

- *Maintaining a biological, social and economic system* considers the health of the human ecosystem as well as the marine ecosystem. A fishery which rotates among multiple species can deplete individual stocks and still be sustainable so long as the ecosystem retains its intrinsic integrity. Such a definition might consider as sustainable fishing practices that lead to the reduction and possible extinction of some species.

Social Sustainability

Fisheries and aquaculture are, directly or indirectly, a source of livelihood for over 500 million people, mostly in developing countries.

Social sustainability can conflict with biodiversity. A fishery is socially sustainable if the fishery ecosystem maintains the ability to deliver products the society can use. Major species shifts within the ecosystem could be acceptable as long as the flow of such products continues. Humans have been operating such regimes for thousands of years, transforming many ecosystems, depleting or driving to extinction many species.

> 66 To a great extent, sustainability is like good art, it is hard to describe but we know it when we see it. 99
>
> — Ray Hilborn,

According to Hilborn, the "loss of some species, and indeed transformation of the ecosystem is not incompatible with sustainable harvests." For example, in recent years, barndoor skates have been caught as bycatch in the western Atlantic. Their numbers have severely declined and they will probably go extinct if these catch rates continue. Even if the barndoor skate goes extinct, changing the ecosystem, there could still be sustainable fishing of other commercial species.

Reconciling Fisheries with Conservation

Management goals might consider the impact of salmon on bear and river ecosystems

At the Fourth World Fisheries Congress in 2004, Daniel Pauly asked, "How can fisheries science and conservation biology achieve a reconciliation?", then answered his own question, "By accepting each other's essentials: that fishing should remain a viable occupation; and that aquatic ecosystems and their biodiversity are allowed to persist."

A relatively new concept is relationship farming. This is a way of operating farms so they restore the food chain in their area. Re-establishing a healthy food chain can result in the farm automatically filtering out impurities from feed water and air, feeding its own food chain, and additionally producing high net yields for harvesting. An example is the large cattle ranch Veta La Palma in southern Spain. Relationship farming was first made popular by Joel Salatin who created a 220 hectare relationship farm featured prominently in Michael Pollan's book *The Omnivore's Dilemma* (2006) and the documentary films, Food, Inc. and Fresh. The basic concept of relationship farming is to put effort into building a healthy food chain, and then the food chain does the hard work.

Overfishing

Overfishing can be sustainable. According to Hilborn, overfishing can be "a misallocation of societies' resources", but it does not necessarily threaten conservation or sustainability".

Overfishing is traditionally defined as harvesting so many fish that the yield is less than it would be if fishing were reduced. For example, Pacific salmon are usually managed by trying to determine how many spawning salmon, called the "escapement", are needed each generation to produce the maximum harvestable surplus. The optimum escapement is that needed to reach that surplus. If the escapement is half the optimum, then normal fishing looks like overfishing. But this is still sustainable fishing, which could continue indefinitely at its reduced stock numbers and yield. There is a wide range of escapement sizes that present no threat that the stock might collapse or that the stock structure might erode.

On the other hand, overfishing can precede severe stock depletion and fishery collapse. Hilborn points out that continuing to exert fishing pressure while production decreases, stock collapses and the fishery fails, is largely "the product of institutional failure."

Today over 70% of fish species are either fully exploited, overexploited, depleted, or recovering from depletion. If overfishing does not decrease, it is predicted that stocks of all species currently commercially fished for will collapse by 2048."

A Hubbert linearization (Hubbert curve) has been applied to the whaling industry, as well as charting the price of caviar, which depends on sturgeon stocks. Another example is North Sea cod. Comparing fisheries and mineral extraction tells us that human pressure on the environment is causing a wide range of resources to go through a Hubbert depletion cycle.

Coastal fishing communities in Bangladesh are vulnerable to flooding from sea-level rises.

Habitat Modification

Nearly all the world's continental shelves, and large areas of continental slopes, underwater ridges, and seamounts, have had heavy bottom trawls and dredges repeatedly dragged over their surfaces. For fifty years, governments and organizations, such as the Asian Development Bank, have encouraged the fishing industry to develop trawler fleets. Repeated bottom trawling and dredging literally flattens diversity in the benthic habitat, radically changing the associated communities.

Changing the Ecosystem Balance

Since 1950, 90 percent of 25 species of big predator fish have gone.

- How we are emptying our seas, *The Sunday Times*, May 10, 2009.

- Pauly, Daniel (2004) Reconciling Fisheries with Conservation: the Challenge of Managing Aquatic Ecosystems Fourth World Fisheries Congress, Vancouver, 2004.

Climate Change

Rising ocean temperatures and ocean acidification are radically altering aquatic ecosystems. Climate change is modifying fish distribution and the productivity of marine and freshwater species. This reduces sustainable catch levels across many habitats, puts pressure on resources needed for aquaculture, on the communities that depend on fisheries, and on the oceans' ability to capture and store carbon (biological pump). Sea level rise puts coastal fishing communities at risk, while changing rainfall patterns and water use impact on inland (freshwater) fisheries and aquaculture.

Ocean Pollution

A recent survey of global ocean health concluded that all parts of the ocean have been impacted by human development and that 41 percent has been fouled with human polluted runoff, overfishing, and other abuses. Pollution is not easy to fix, because pollution sources are so dispersed, and are built into the economic systems we depend on.

The United Nations Environment Programme (UNEP) mapped the impacts of stressors such as climate change, pollution, exotic species, and over-exploitation of resources on the oceans. The report shows at least 75 percent of the world's key fishing grounds may be affected.

Diseases and Toxins

Large predator fish contain significant amounts of mercury, a neurotoxin which can affect fetal development, memory, mental focus, and produce tremors.

Irrigation

Lakes are dependent on the inflow of water from its drainage basin. In some areas, aggressive irrigation has caused this inflow to decrease significantly, causing water depletion and a shrinking of the lake. The most notable example is the Aral Sea, formerly among the four largest lakes in the world, now only a tenth of its former surface area.

Abandoned ship near Aral, Kazakhstan.

Remediation

Fisheries Management

Fisheries management draws on fisheries science to enable sustainable exploitation. Modern fisheries management is often defined as mandatory rules based on concrete objectives and a mix of management techniques, enforced by a monitoring control and surveillance system.

- Ideas and rules: Economist Paul Romer believes sustainable growth is possible providing the right ideas (technology) are combined with the right rules, rather than simply hectoring fishers. There has been no lack of innovative ideas about how to harvest fish. He characterizes failures as primarily failures to apply appropriate rules.

- Fishing subsidies: Government subsidies influence many of the world fisheries. Operating cost subsidies allow European and Asian fishing fleets to fish in distant waters, such as West Africa. Many experts reject fishing subsidies and advocate restructuring incentives globally to help struggling fisheries recover.

- Economics: Another focus of conservationists is on curtailing detrimental human activities by improving fisheries' market structure with techniques such as salable fishing quotas, like those set up by the Northwest Atlantic Fisheries Organization, or laws.

- Payment for Ecosystem Services: Environmental Economist, Essam Y Mohammed, argues that by creating direct economic incentives, whereby people are able to receive payment for the services their property provides, will help to establish sustainable fisheries around the world as well as inspire conservation where it otherwise would not.

- Sustainable fisheries certification: A promising direction is the independent certification programs for sustainable fisheries conducted by organizations such as the Marine Stewardship Council and Friend of the Sea. These programs work at raising consumer awareness and insight into the nature of their seafood purchases.

- Ecosystem based fisheries.

Ecosystem Based Fisheries

"We propose that rebuilding ecosystems, and not sustainability per se, should be the goal of fishery management. Sustainability is a deceptive goal because human harvesting of fish leads to a progressive simplification of ecosystems in favour of smaller, high turnover, lower trophic level fish species that are adapted to withstand disturbance and habitat degradation."

Tony Pitcher and Daniel Pauly,

According to marine ecologist Chris Frid, the fishing industry points to marine pollution and global warming as the causes of recent, unprecedented declines in fish populations. Frid counters that overfishing has also altered the way the ecosystem works. "Everybody would like to see the rebuilding of fish stocks and this can only be achieved if we understand all of the influences, human and natural, on fish dynamics." He adds: "fish communities can be altered in a number of ways, for example they can decrease if particular-sized individuals of a species are targeted, as this affects predator and prey dynamics. Fishing, however, is not the sole cause of changes to marine life—pollution is another example... No one factor operates in isolation and components of the ecosystem respond differently to each individual factor."

The traditional approach to fisheries science and management has been to focus on a single species. This can be contrasted with the ecosystem-based approach. Ecosystem-based fishery concepts have been implemented in some regions. In a 2007 effort to "stimulate much needed discussion" and "clarify the essential components" of ecosystem-based fisheries science, a group of scientists offered the following ten commandments for ecosystem-based fisheries scientists

- Keep a perspective that is holistic, risk-adverse and adaptive.

- Maintain an "old growth" structure in fish populations, since big, old and fat female fish have been shown to be the best spawners, but are also susceptible to overfishing.

- Characterize and maintain the natural spatial structure of fish stocks, so that management boundaries match natural boundaries in the sea.

- Monitor and maintain seafloor habitats to make sure fish have food and shelter.

- Maintain resilient ecosystems that are able to withstand occasional shocks.

- Identify and maintain critical food-web connections, including predators and forage species.

- Adapt to ecosystem changes through time, both short-term and on longer cycles of decades or centuries, including global climate change.

- Account for evolutionary changes caused by fishing, which tends to remove large, older fish.

- Include the actions of humans and their social and economic systems in all ecological equations.

Marine Protected Areas

Strategies and techniques for marine conservation tend to combine theoretical disciplines, such as population biology, with practical conservation strategies, such as setting up protected areas, as with Marine Protected Areas (MPAs) or Voluntary Marine Conservation Areas. Each nation defines MPAs independently, but they commonly involve increased protection for the area from fishing and other threats.

Marine life is not evenly distributed in the oceans. Most of the really valuable ecosystems are in relatively shallow coastal waters, above or near the continental shelf, where the sunlit waters are often nutrient rich from land runoff or upwellings at the continental edge, allowing photosynthesis, which energizes the lowest trophic levels. In the 1970s, for reasons more to do with oil drilling than with fishing, the U.S. extended its jurisdiction, then 12 miles from the coast, to 200 miles. This made huge shelf areas part of its territory. Other nations followed, extending national control to what became known as the exclusive economic zone (EEZ). This move has had many implications for fisheries conservation, since it means that most of the most productive maritime ecosystems are now under national jurisdictions, opening possibilities for protecting these ecosystems by passing appropriate laws.

Daniel Pauly characterises marine protected areas as "a conservation tool of revolutionary importance that is being incorporated into the fisheries mainstream." The Pew Charitable Trusts have funded various initiatives aimed at encouraging the development of MPAs and other ocean conservation measures.

Fish Farming

There exists concerns that farmed fish cannot produce necessary yields efficiently. For example, farmed salmon eat three pounds of wild fish to produce one pound of salmon.

Laws and Treaties

International laws and treaties related to marine conservation include the 1966 Convention on Fishing and Conservation of Living Resources of the High Seas. United States laws related to marine conservation include the 1972 Marine Mammal Protection Act, as well as the 1972 Marine Protection, Research and Sanctuaries Act which established the National Marine Sanctuaries program. Magnuson-Stevens Fishery Conservation and Management Act.

Awareness Campaigns

Introducing the results of long term monitoring to a local fishermen in Kihnu, Estonia.

Various organizations promote sustainable fishing strategies, educate the public and stakeholders, and lobby for conservation law and policy. The list includes the Marine Conservation Biology Institute and Blue Frontier Campaign in the U.S., The U.K.'s Frontier (the Society for Environmental Exploration) and Marine Conservation Society, Australian Marine Conservation Society, International Council for the Exploration of the Sea (ICES), Langkawi Declaration, Oceana, PROFISH, and the Sea Around Us Project, International Collective in Support of Fishworkers, World Forum of Fish Harvesters and Fish Workers, Frozen at Sea Fillets Association and CEDO.

The United Nations Millennium Development Goals include, as goal #7: target 2, the intention to "reduce biodiversity loss, achieving, by 2010, a significant reduction in the rate of loss", including improving fisheries management to reduce depletion of fish stocks.

Some organizations certify fishing industry players for sustainable or good practices, such as the Marine Stewardship Council and Friend of the Sea.

Other organizations offer advice to members of the public who eat with an eye to sustainability. According to the marine conservation biologist Callum Roberts, four criteria apply when choosing seafood:

- Is the species in trouble in the wild where the animals were caught?

- Does fishing for the species damage ocean habitats?

- Is there a large amount of bycatch taken with the target species?

- Does the fishery have a problem with discards—generally, undersized animals caught and thrown away because their market value is low?

The following organizations have download links for wallet-sized cards, listing good and bad choices:

- Monterey Bay Aquarium Seafood Watch, USA

- Blue Ocean Institute, USA

- Marine Conservation Society, UK

- Australian Marine Conservation Society

- The Southern African Sustainable Seafood Initiative

Data Issues

Data Quality

One of the major impediments to the rational control of marine resources is inadequate data. According to fisheries scientist Milo Adkison (2007), the primary limitation in fisheries management decisions is poor data. Fisheries management decisions are often based on population models, but the models need quality data to be accurate. Scientists and fishery managers would be better served with simpler models and improved data.

Unreported Fishing

Estimates of illegal catch losses range between $10 billion and $23 billion annually, representing between 11 and 26 million tonnes.

- Incidental catch

Shifting Baselines

Shifting baselines is a term which describes the way significant changes to a system are measured against previous baselines, which themselves may represent significant changes from the original state of the system. The term was first used by the fisheries scientist Daniel Pauly in his paper "Anecdotes and the shifting baseline syndrome of fisheries". Pauly developed the term in reference to fisheries management where fisheries scientists sometimes fail to identify the correct "baseline" population size (e.g. how abundant a fish species population was *before* human exploitation) and thus work with a shifted baseline. He describes the way that radically depleted fisheries were evaluated by experts who used the state of the fishery at the start of their careers as the baseline, rather than the fishery in its untouched state. Areas that swarmed with a particular species hundreds of years ago, may have experienced long term decline, but it is the level of decades previously that is considered the appropriate reference point for current populations. In this way large declines in ecosystems or species over long periods of time were, and are, masked. There is a loss of perception of change that occurs when each generation redefines what is "natural".

Looting the Seas

Looting the seas is the name given by the International Consortium of Investigative Journalists to a series of journalistic investigations into areas directly affecting the sustainability of fisheries. So far they have investigated three areas involving fraud, negligence and overfishing:

- The black market in bluefin tuna

- Subsidies propping up the Spanish fishing industry

- Overfishing of the southern jack mackerel

Other Factors

The focus of sustainable fishing is often on the fish. Other factors are sometimes included in the broader question of sustainability. The use of non-renewable resources is not fully sustainable. This might include diesel fuel for the fishing ships and boats: there is even a debate about the long term sustainability of biofuels. Modern fishing nets are usually made of artificial polyamides like nylon. Synthetic braided ropes are generally made from nylon, polyester, polypropylene or high performance fibers such as high modulus polyethylene (HMPE) and aramid.

Energy and resources are employed in fish processing, refrigeration, packaging, logistics, etc. The methodologies of Life-cycle assessment are useful to evaluate the sustainability of components and systems. These are part of the broad question of sustainability.

Fishery Regulations

- Set catch limits well below the maximum sustainable yield

- Improve monitoring and enforcement of regulations

- Economic Approaches

- Sharply reduce or eliminate fishing subsidies

- Charge fees for harvesting fish and shellfish from publicly owned offshore waters

- Certify sustainable fisheries

- Protected Areas

- Establish no-fishing areas

- Establish more marine protected areas

- Rely more on integrated coastal management

- Consumer Information

- Label sustainably harvested fish

- Publicize overfished and threatened species

Bycatch

- Use wide-meshed nets to allow escape of smaller fish

- Use net escape devices for sea birds and sea turtles

- Ban throwing edible and marketable fish back into the sea

Aquaculture

- Restrict coastal locations for fish farms

- Control pollution more strictly

- Depend more on herbivorous fish species

Nonnative Invasions

- Kill organisms in ship ballast

- water Filter organisms from ship ballast water

- Dump ballast water far at sea and replace with deep-sea water

Protecting Wetlands

- Legally protect existing wetlands

- Steer development away from existing wetlands

- Use mitigation banking only as a last resort

- Require creation and evaluation of a new wetland before destroying an existing wetland

- Restore degraded wetlands

- Try to prevent and control invasions by nonnative species

Capture Fisheries

Aquaculture has been defined in many ways. It has been called as the rearing of aquatic organisms under controlled or semi controlled condition - thus it is underwater agriculture. The other definition of aquaculture is the art of cultivating the natural produce of water, the raising or fattening of fish in enclosed ponds. Another one is simply the large-scale husbandry or rearing of aquatic organisms for commercial purposes. Aquaculture can be a potential means of reducing over need to import fishery products, it can mean an increased number of jobs, enhanced sport and commercial fishing and a reliable source of protein for the future.

Fish is a rich source of animal protein and its culture is an efficient protein food production system from aquatic environment. The main role of fish culture is its contribution in improving the nutritional standards of the people. Fish culture also helps in utilising water and land resources. It provides inducement to establish other subsidiary industries in the country.

Fisheries can be categorised into two types - fin fisheries and non-fin fisheries. The former is fisheries of true fishes, whereas the latter is the fisheries of organisms other than true fish like prawn, crab, lobster, mussel, oyster, sea cucumbers, frog, sea weeds, etc.

Fin fisheries can be further categorized into two types – capture fisheries and culture fisheries.

Capture fisheries is exploitation of aquatic organisms without stocking the seed. Recruitment of the species occurs naturally. This is carried out in the sea, rivers, reservoirs, etc. Fish yield decreases gradually in capture fisheries due to indiscriminate catching of fish including brooders and juveniles. Overfishing destroys the fish stocks. Pollution and environmental factors influence the fish yield. The catches include both desirable and undesirable varieties.

A culture fishery is the cultivation of selected fishes in confined areas with utmost care to get maximum yield. The seed is stocked, nursed and reared in confined waters, then the crop is harvested. Culture takes place in ponds, which are fertilized and supplementary feeds are provided to fish to get maximum yield. In order to overcome the problems found in capture fisheries to increase the production, considerable attention is being given to the culture fisheries.

Culture fisheries are conducted in freshwater, brackish water and sea waters. With the development and expansion of new culture systems, farming of a wide variety of aquatic organisms like prawns, crabs, molluscs, frogs, sea weeds, etc. have come under culture fisheries. Due to the culture of a variety of aquatic organisms, culture fisheries has been termed as aquaculture.

Capture fisheries is intended for catching fishes and also prawns, lobsters, crabs, sea-cucumbers, whales, pearl oysters, edible bivalve and copious other organisms of other than fishes etc. Primitive human beings were acquainted with capture fishery centuries passed for him to observe and understand for the possibilities of culturing fish. Then also he depended mostly on the culture of fishes with parental care. Later, he tried to collect the fingerlings in canals, distribution canals. In the earlier days, the mixture of carnivore fish fingerlings and carp fish fingerlings were stocked together in tanks. Later, they were segregated and stocked selecting the required variety.

Capture of fishes can be broadly divided in to two types:

a) Capture by Human effort

b) Capture by observing the behavioural pattern of Fishes.

Inland capture fishery of India has an important place; it contributes to about 30% of the total fish production. The large network of inland water masses will continue to provide great potential, for economic capture fishery which consequently will compete well with fast growing fish-culture practices. The freshwater inland water bodies fall into five major categories, distinguished as the Ganga, the Brahmaputra and the Indus system of the Northern India, and the East and the West coast river systems of the Southern (peninsular) India. These river systems have certain characteristics of their own with respect to their ecology, climatic conditions and fish populations of commercial food fishes. Besides, there are a number of land-locked lakes especially those situated at high altitudes which have started supporting cold water fisheries of both indigenous and exotic species. In addition to the above-mentioned freshwaters, there are also rich fisheries offered by extensive brackish waters, including important estuaries (Hooghly - Matlah, Mahanadi and Godavari estuaries), lagoons (Chilka lake, Pulicat lake) and backwaters (Vembanad) and paddy fields (Pokkali in Kerala). Chilka lake in the state of Orissa is an open shallow brackish water lake having an area of 906 sq. km. in summer and 1165 sq. km. in rainy season. A long canal joins it with sea. Waters from river Daya (Mahanadi) and other smaller streams flow into it. Recent additions to the natural inland water bodies are man-made reservoirs. There are at present some 300 reservoirs which hold very good prospects, after restocking, both for capture as well as for culture fisheries. Some of these reservoirs have responded fairly well to attempts to restock them with indigenous as well as exotic species.

Inland capture fishery is a continually expanding industry bringing under its fold newer fisheries of a local or regional nature, while improving upon those which are existing already. Introduction of exotic species from abroad and inter-regional transplantation of fish from Northern to Southern waters have been most welcome and rewarding.

The inland capture fishery, however, stands at a critical juncture, which draws a special attention at the national level. Rapid industrialization movements in the country have given a serious blow to the growth of the inland fisheries which was struggling to come out of the old-fashioned style to a more rational and scientific style. Constructions of dams have been the cause of decline and damage to several regionally important fisheries. Discharge from industrial establishments, multiplying at mushroom growth, into inland water bodies is polluting the water in very serious proportions, and is damaging the fish populations tremendously. Already, old-age practices of indiscriminate fishing of fingerlings and juveniles, supporting local and seasonal fisheries, especially in breeding or nursery grounds, have been doing enormous damage, and needed effective controls for conservation. Likewise, time-old practice of sewage disposal into rivers was a menacing practice causing heavy pollution. Great harm is also being done from agricultural wash coming to inland waters, which brings to fish a very toxic principle of the numerous pesticides used in the agricultural practices.

Conservation Strategies

Due to factors such as human modifications to the environment, overexploitation, habitat loss, exotic species and others, aquatic biodiversity is greatly threatened. Ecosystems and species important in sustaining human life and the health of the environment are disappearing at an alarming rate. In order to preserve these threatened areas and species for future generations, immediate action in the form of aquatic biodiversity conservation strategies are necessary.

In general, aquatic conservation strategies should support sustainable development by protecting biological resources in ways that will preserve habitats and ecosystems. In order for biodiversity conservation to be effective, management measures must be broad based. This can be achieved through many mechanisms including:

- Aquatic Diversity Management Areas (ADMAs): As first proposed by Moyle and Yoshiyama (1994) the creation of ADMAs, are a systematic management approach for watersheds, where the primary goal is to protect the aquatic biodiversity in a given area. ADMAs range from individual species protection acts to full-scale biodiversity oriented programs. The best way to properly manage ADMAs is to stop or greatly reduce all human activity contributing to habitat degradation in that area. This concept has been applied in the Sierra Nevada area.

- Marine Reserves: A marine reserve is a defined space within the sea in which fishing is banned or other restrictions are placed in an effort to protect plants, animals, and habitats, ultimately conserving biodiversity. Marine reserves can also be used for educational purposes, recreation, and tourism as well as potentially increasing fisheries yields by enhancing the declining fish populations. Marine reserves are also very similar to marine protected areas, fishery reserves, sanctuaries, and parks.

- Bioregional Management: Bioregional management is a total ecosystem strategy, which regulates factors affecting aquatic biodiversity by balancing conservation, economic, and social needs within an area. This consists of both small-scale biosphere reserves and larger reserves. Biosphere reserves, generally small in scale, have a strong conservation focus, and consist of one or more protected central habitats and surrounding buffer zones. In these bioresevation units, activities such as fishing, hunting, harvesting, and development activities are strictly limited. In contrast, nonbiosphere reserve areas encompass much broader ranges, and many more habitat types (e.g., the Florida Keys National Marine Sanctuary). Other examples of National Marine Sanctuaries include Stellwagen bank, and Monterey Bay.

- Threatened or endangered species designations: The World Resources Institute documents that the designation of a particular species as threatened or endangered has historically been the primary method of protecting freshwater biodiversity. Threatened species include organisms likely to become endangered if not properly protected. Endangered species are plants and animals that need protection in order to survive, as they are in immediate danger of becoming extinct. Once species are "listed," they become subject to national recovery programs and will be placed under international protection. Severe monetary penalties can occur if threatened and endangered species regulations are broken, and can even result in jail sentences. For more information, please visit EPA's Endangered Species Protection Program, Endangered Species Act, or U.S. Fish and Wildlife Service's Endangered Species Program.

- Local watershed groups: Rivers and streams, regardless of their condition, often go unprotected since they often pass through more than one political jurisdiction, making it difficult to enforce conservation and management of resources. However, in recent years, the protection of lakes and small portions of watersheds organized by local watershed groups has helped this situation.

- Freshwater Initiatives: The Nature Conservancy has instituted a program referred to as the Freshwater Initiative (FWI). The objective of the FWI is to significantly increase freshwater conservation within the United States and other areas, through three strategies: watershed action, water science, and water lessons.

- Specialized Programs: Many specialized programs have been instituted to protect biodiversity. For example, the USDA Forest Service initiated Bring Back the Natives, a cooperative state-federal program. The goal of this program is to restore the health of riverine systems and associated species. Areas targeted for this program include lands managed by the U.S. Forest Service and the Bureau of Land management.

- Research: Various organizations and conferences that research biodiversity and associated conservation strategies help to identify areas of future research analyze current trends in aquatic biodiversity, even conduct specialized studies. Examples of such organizations include the Nature Conservancy, Natural Heritage Network, World Conservation Monitoring Centre, World Resources Institute, NOAA Fisheries Office of Protected Resources , and Convention on Biological Diversity (CBD).

- Increase Public Awareness: Increasing public awareness is one of the most important ways to conserve aquatic biodiversity. This can be accomplished through educational programs, incentive programs, and volunteer monitoring programs. For example, the State of Delaware has an Adopt-a-Wetland Program, designed to increase public awareness as to the value and of wetlands and the need for conservation.

- Restoration/Mitigation Efforts: Aquatic areas that have been damaged or suffered habitat loss or degradation can be restored. Even species populations that have suffered a decline can be targeted for restoration (e.g., Pacific Northwest salmon populations). Some management practices such as the establishment of riparian buffer zones and the restoration of natural flow patterns and discharge regimes are being applied to riverine areas. Recently, habitat restoration has also been performed in various areas to replace losses from dredging projects and in many wetland habitats.

- Regulatory Measures: This may include wastewater discharge regulations like NPDES or fishery conservation measures, fisheries management councils, even fishery bans. For example, the Magnuson-Stevens Fishery Conservation and Management Act of 1976 and the associated 1996 Sustainable Fisheries Amendment require the conservation and management of the marine fishery resources in the United States, predominately managed by NOAA and National Marine Fisheries Service (NMFS). This creation of sustainable fisheries is largely completed through regulatory actions including the collection of the best scientific data available.

- Local community actions: The demand for freshwater - and the threats to its health - originate from the actions of millions of people. To solve these challenges also requires actions of many. State and federal governments, and many local governments and public agencies, are already at work. So, too, are numerous citizen volunteers. Any individual can take steps to make healthy water a welcome part of everyday life.

References

- Ray, G. Carleton (2004) "Issues and Mechanisms", part 1 in Coastal-marine Conservation: Science and Policy. Malden, MA: Blackwell Pub. ISBN 978-0-632-05537-1

- Jenkins, Lekeliad (2010). "Profile and influence of the successful fisher-Inventor of marine conservation technology". Conservation and Society. 8: 44. doi:10.4103/0972-4923.62677

- Carleton, Ray G.; McCormick, Jerry (1 April 2009). Coastal-Marine Conservation: Science and Policy. John Wiley & Sons. ISBN 978-1-4443-1124-2

- Gell, FR; Roberts, CM (2003). "Benefits beyond boundaries: the fishery effects of marine reserves". Trends in Ecology & Evolution. 18 (9): 448–455. doi:10.1016/S0169-5347(03)00189-7

- Thys, Tierney (30 November 2003). "Tracking Ocean Sunfish, Mola mola with Pop-Up Satellite Archival Tags in California Waters". OceanSunfish.org. Retrieved 14 June 2007

- Bodeo-Lomicky, Aidan (4 February 2015). The Vaquita: TheBiology of an Endangered Porpoise. Createspace Independent Pub. ISBN 978-1-5077-5577-8

- Pauly, D; Christensen, V; Guénette, S; Pitcher, TJ; Sumaila, UR; Walters, CJ; Watson, R; Zeller, D (2002). "Towards sustainability in world fisheries". Nature. 418: 689–695. doi:10.1038/nature01017

- Norse, Elliott A. and Crowder, Larry B. (Eds.) (2005) Marine Conservation Biology: The Science of Maintaining the Sea's Biodiversity, Island Press. ISBN 978-1-55963-662-9

- Hughes, TP; Bellwooda, DR; Folkeb, C; Steneck, RS; Wilson, J (2005). "New paradigms for supporting the resilience of marine ecosystems". Trends in Ecology & Evolution. 20 (7): 380–386. doi:10.1016/j.tree.2005.03.022

Permissions

We would like to thank the editorial team for lending their expertise to make the book truly unique. They have played a crucial role in the development of this book. Without their invaluable contributions this book wouldn't have been possible. They have made vital efforts to compile up to date information on the varied aspects of this subject to make this book a valuable addition to the collection of many professionals and students.

This book was conceptualized with the vision of imparting up-to-date and integrated information in this field. To ensure the same, a matchless editorial board was set up. Every individual on the board went through rigorous rounds of assessment to prove their worth. After which they invested a large part of their time researching and compiling the most relevant data for our readers.

The editorial board has been involved in producing this book since its inception. They have spent rigorous hours researching and exploring the diverse topics which have resulted in the successful publishing of this book. They have passed on their knowledge of decades through this book. To expedite this challenging task, the publisher supported the team at every step. A small team of assistant editors was also appointed to further simplify the editing procedure and attain best results for the readers.

Apart from the editorial board, the designing team has also invested a significant amount of their time in understanding the subject and creating the most relevant covers. They scrutinized every image to scout for the most suitable representation of the subject and create an appropriate cover for the book.

The publishing team has been an ardent support to the editorial, designing and production team. Their endless efforts to recruit the best for this project, has resulted in the accomplishment of this book. They are a veteran in the field of academics and their pool of knowledge is as vast as their experience in printing. Their expertise and guidance has proved useful at every step. Their uncompromising quality standards have made this book an exceptional effort. Their encouragement from time to time has been an inspiration for everyone.

The publisher and the editorial board hope that this book will prove to be a valuable piece of knowledge for students, practitioners and scholars across the globe.

Index

www.ingramcontent.com/pod-product-compliance
Lightning Source LLC
Chambersburg PA
CBHW082018190326
41458CB00010B/3225